Eduardo Pagel Floriano

DENDROMETRY

Eduardo Pagel Floriano

DENDROMETRY

Measuring trees and their growth

ScienciaScripts

Imprint

Any brand names and product names mentioned in this book are subject to trademark, brand or patent protection and are trademarks or registered trademarks of their respective holders. The use of brand names, product names, common names, trade names, product descriptions etc. even without a particular marking in this work is in no way to be construed to mean that such names may be regarded as unrestricted in respect of trademark and brand protection legislation and could thus be used by anyone.

Cover image: www.ingimage.com

This book is a translation from the original published under ISBN 978-620-6-76223-2.

Publisher:
Sciencia Scripts
is a trademark of
Dodo Books Indian Ocean Ltd. and OmniScriptum S.R.L publishing group

120 High Road, East Finchley, London, N2 9ED, United Kingdom
Str. Armeneasca 28/1, office 1, Chisinau MD-2012, Republic of Moldova, Europe
Printed at: see last page
ISBN: 978-620-8-27321-7

EDUARDO PAGEL FLORIANO

DENDROMETRY

Rio Largo, AL, Brazil

2024

SUMMARY

PRESENTATION

This compendium is another response to requests for bibliography from students on the Forestry Engineering course at the Federal University of Alagoas' Engineering and Agrarian Sciences Campus.

We have tried to bring together most of the activities and techniques related to dendrometry in this work, so that it can serve as a reference for teaching this subject in our course.

Maceió, September 15, 2024.

Eduardo Pagel Floriano

1 INTRODUCTION

Dendrometry is the science that studies the quantitative measurement and qualitative evaluation of trees and their direct products. The aim of dendrometry is to obtain data on the different dendrometric variables and attributes with regard to their use and conservation and to provide information for their proper management. The objects of measurement in dendrometry are trees or parts of them, samples (made up of sampling units) and the products obtained directly from the trees, or obtained from their primary transformation. Dendrometry uses appropriate equipment and measurement methods to obtain values for tree variables and attributes, which are processed, summarized and evaluated according to the purposes of the measurements, using mathematical and statistical techniques.

1.1 Definitions

1.1.1 Tree

A tree is a woody plant made up of a trunk, crown and roots, which reaches a height of 5 meters or more at maturity.

Woody plant, usually tall, with an erect main stem, or trunk, fixed to the ground with roots, and branching into leaf-laden branches that form the canopy (DICIONÁRIO MICHAELIS ON LINE, 2021).

1.1.2 Beam

The stem is the commercial part of the trunk considering the standing tree, i.e. including the stump if it is felled. In sympodial trees, it extends from the base near the ground to the point where the branches of the crown are inserted. In trees with monopodial growth, it extends to a minimum diameter of the trunk.

1.1.3 Dendrometry

Dendrometry is the science of measuring trees.

The word comes from the Greek: Dendro (tree) + Metria (measurement). However, the subject involves more than this, dealing with measurements of trees in stands using sample plots to obtain diametric distributions, tree density, estimates of stocks per hectare, the growth of trees and forest stands in their different dimensions and the study of the relationships between dendrometric variables, allowing one to be estimated as a function of the other.

Other definitions of dendrometry can be found in the literature, such as:

> "Dendrometry is the branch of forestry science that deals with determining and/or estimating the dimensions of trees, stands and forests, their growth and their products." (IMAÑA et al, 2002).

> "Dendrometry is a branch of Forest Science that deals with the determination or estimation of forest resources, either of the tree itself or of the stand itself, with the aim of accurately predicting the volume, increment or production of a given forest resource." (SILVA et al, 1979).

1.1.4 Forest

Forests are lands covering more than 0.5 hectares with trees over 5 meters tall and a canopy cover of more than 10%, or trees capable of reaching these thresholds in situ (FAO, 2015).

1.1.5 Dendrometric variables

These are the values of trees and their parts that can be measured or counted.

Measured dendrometric variables are continuous and take on real values, for example: height. Variables that can only be counted are discrete, taking on integer values; example: number of seeds per kilogram.

The main dendrometric variables are the diameter (d) or circumference (c) of the trunk taken at a height of 1.3 m, the height (h) of the tree measured from its base to its top, the individual basal area (g) corresponding to the diameter of the tree (Figura 1), individual trunk volume (v) and age (t).

Obtaining statistics on forest stands is the subject of the "Forest Inventory" course.

1.1.6 Tree attributes

It is the qualitative data of the trees that distinguishes them from each other. Attributes are usually distributed into quality classes to enable them to be analyzed. An example of this is the quality classes of sawn timber:

> Class 1 - wood free of knots or imperfections;

> Class 2 - wood with few live knots adhered to it, or with minimal imperfections;
> Class 3 - wood with dry knots adhered to it, or with small imperfections;
> Class 4 - wood with dry, loose knots or many imperfections.

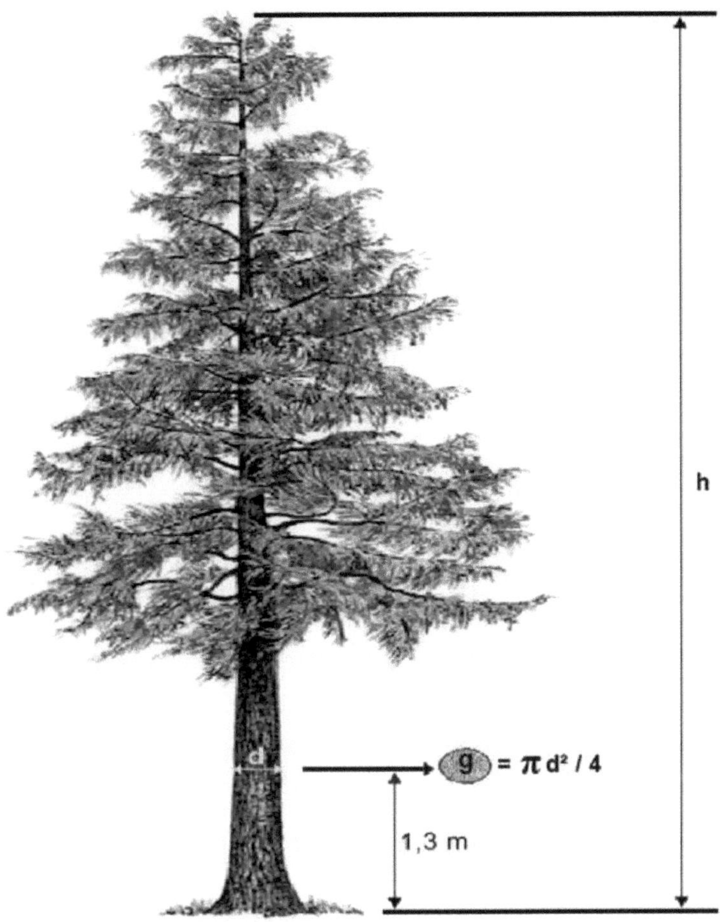

Figura 1 - Diameter (d), height (h) and basal area (g). Image: modified from Bourdo (2001).

1.1.7 Measurement precision and accuracy

Precision is represented by the amount of variation in the results of a repeated set of measurements of the same thing. For example: measuring the height of the same tree 5 times, 22.7 m,

13

22.5 m, 22.6 m, 22.5 m and 22.4 m were obtained, giving an average value of 22.54 m, with a variation of 22.4 m to 22.7 m.

Accuracy refers to the deviation of the sample from the true parameter including measurement errors and other non-sampling errors, which is difficult to obtain. Example: in the same sample as in the previous paragraph, the tree was felled and the precise and true height was measured as 22.63 m, i.e. the error of the average in relation to the true value was -0.09 m, which represents around -0.4% error, or 99.6% accuracy.

1.1.8 Sampling and sampling unit

1.1.8.1 Sample

A sample is a small part of a population chosen to represent it. In forestry, because populations are usually very large, sampling is used to measure a few individuals concentrated in sample units, also called plots, and these plots are distributed over the population according to appropriate statistical schemes. The sample is the entire set of sample units measured in the population.

1.1.8.2 Sample units

Forest sampling units are any parts of the population that are intended to represent it and that contain individuals or parts of individuals from the population. A sampling unit can be just one tree, or a plot containing a group of trees, or even parts of trees. A sampling unit may or may not have a fixed area. It can be, for example: a forest area of 600 m², 20 m wide by 30 m long;

it can also be a single tree; or a set containing a fixed number of 6 trees, for example.

To represent a population, the sampling units must be distributed over the population according to statistical schemes, which are the subject of the "Forest Inventory" course.

1.1.8.3 Population parameters

When you measure all the individuals in a population and calculate their average, this average is a population parameter, it is the true average of the population.

1.1.8.4 Sample statistics

When you measure a sample of some individuals from a population and calculate their mean, this mean is a statistic by which you estimate the population parameter; it is a mean that is close to the true population mean with a certain probability of error.

1.2 Importance of Dendrometry

All of the Forest Engineer's work is based on measurements and assessments of trees and forest stands, their growth and evolution. It can be said that dendrometry is the foundation of forest engineering. Dendrometry is the basis for disciplines such as forest inventory and forest management. Knowing how to correctly measure trees, their growth and evolution and assess their quality is essential for forest engineers and foresters. The main relationships between dendrometry and other disciplines are shown in Figura 2.

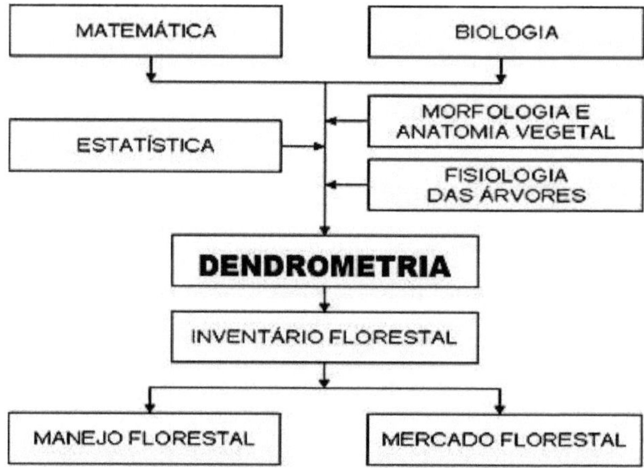

Figura 2 - The relationship between dendrometry and other disciplines.

1.3 History

Some historical milestones in the evolution of dendrometry as a science are listed sequentially below:

> Until the middle of the 18th century, forest measurements were taken by eye;

> From 1750 onwards, tapes were used to measure diameters and auxiliary instruments to measure heights;

> 1759 - Doebel and Beckmann presented studies to determine the volume of wood in forest stands;

- 1764 - DUHAMEL de MONCEAU published "Da exploração da floresta" with references to dendrometry.
- 1765 - Oettel considered the trunk to be a geometric figure and calculated mathematical correlations between dendrometric variables;
- 1787 - Paulsen published the first volume tables, followed by Cotta in 1804;
- 1791 - Hennert demonstrated xylometric measurement;
- 1794 - KAESTNERT introduced the determination of the volume of felled trees;
- History
- 1804 - Heinrich Cotta made an instrument to measure diameters;
- 1812 - Hossheld introduced the first procedures for sampling and evaluating forest stands;
- 1825 - Huber developed his rigorous cubing method;
- 1837 - Smalian described his cubing formula;
- It was only in the first half of the 20th century that forest measurements became more precise and evolved to the current situation;
- 1948 - Bitterlich introduced the angular exclusion measurement method;
- Since then, the works of HOHENADL (1856-1950); SPURR (1952); KRENN (1908-1948); ASSMANN (1961); PRODAN (1965); among other forest scientists, have stood out.

2 DENDROMETRIC VARIABLES

The standard "TB15: The Standardization of Symbols in Forest Mensuration" issued by the *International Union of Forestry Research Organizations* (IUFRO) standardized the international symbology used for dendrometric variables in 1959. According to IUFRO, when making the recommendations on the standardization of symbols, several conflicting interests had to be considered and reconciled, the most important of which were:

> "1. Symbols should be easy to remember; they should be simple and there should not be too many of them;
>
> 2. they must be easy to reproduce on typewriters and in print;
>
> 3. must not conflict with mathematical or other symbols commonly found in forestry literature;
>
> 4. symbols that have already become well-established internationally should not be changed, if possible;
>
> 5. the symbols must have precise meanings." (IUFRO, 1959)

In this work, we tried to follow the international standard TB15. The main dendrometric variables are described below.

The individual dendrometric variables of the trees are expressed by lowercase letters, such as d, h, g, v and i, meaning diameter, height, basal area, volume and increment in one year, respectively, in relation to individual trees. Individual averages are also represented by lowercase letters, but overlaid by a bar, as in $\bar{d}$ which stands for average diameter. The collective variables per hectare are expressed with capital letters such as N, G, V and I, meaning number of trees per hectare, basal area per hectare, volume per hectare and increment per hectare in one year, respectively.

2.1 Circumference or perimeter (c)

This is the measurement of the perimeter of the torso using a tape measure (Figura 3).

The measurement of the circumference at breast height (c), taken at 1.3 m height, corresponds to the value of π multiplied by the diameter (d) of the tree at 1.3 m height. The mnemonic CAP (Circumference at Breast Height) is also used to represent it. When taken at a height of 1.3 m, it can be represented simply by c.

It is measured with measuring tapes with a precision of 1 millimeter, 5 millimeters or 1 centimeter, the former being preferred due to its greater precision (Figura 4).

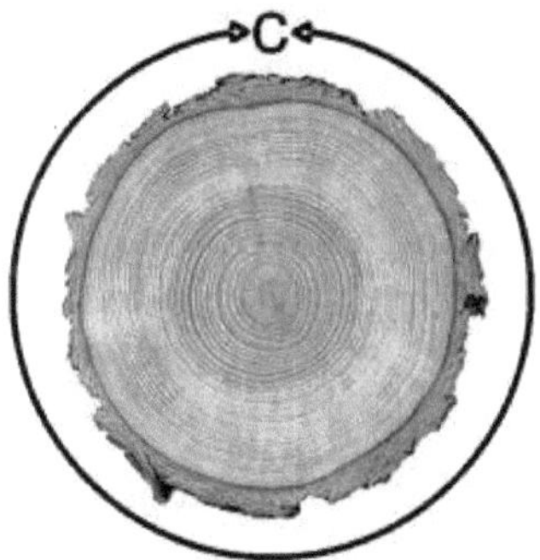

Figura 3 - Trunk circumference.

Trunk circumference measurements taken at different heights can be represented by c_i , where i is the height at which the circumference was taken. For example, a trunk circumference taken at a height of 2.5 m can be represented by $c_{.2,5}$

Similarly, circumferences at a height of 1.3 m taken at different ages can be represented by c_t , where t is the age of the tree when it was measured. For example, the circumference of a tree measured at 12 years of age can be represented by $c_{.12}$

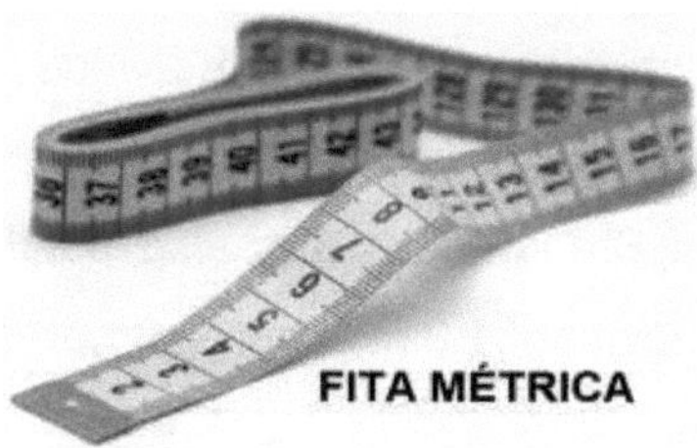

Figura 4 - Tape measure for measuring circumferences.

The unit of measurement for circumferences is the centimeter and the number of decimal places should not be less than or exceed the precision of the equipment used for measurement, except when representing averages, when it is recommended to use one decimal place more than that of the measuring equipment. Tapes do not usually have a precision of more than 1 millimeter, so the circumference value should be presented as 36.4 cm, while the average of 36.4 and 36.1 should be presented as 36.25 cm.

When the circumferences (c) are measured at a height of 1.3 m, the diameters (d) are obtained by dividing the circumference measurement by the value of pi ($\pi = 3.141593$).

2.2 Diameter

2.2.1 Diameter at Breast Height (d)

The diameter (d) of the trees is measured at 1.3 m above the ground and can also be represented by the mnemonic DAP (Diameter at Breast Height), measured using a diametric tape or suta (Figura 5); with a tape it is obtained by a single measurement or by two perpendicular measurements.

Figura 5 - Electronic forestry probe for measuring diameters. Source: TerraGes, 2021.

The diametric tapes (Figura 6) have their measurements multiplied by π, so the diameters are obtained directly, without the need for transformation.

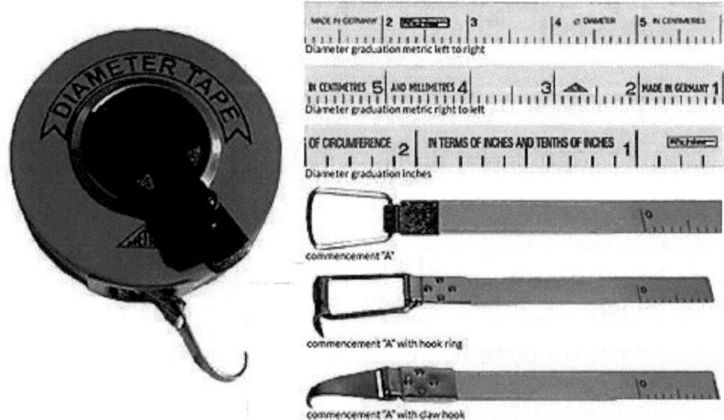

Figura 6 - Diametric tape. Source: TerraGes, 2021.

It is common knowledge that two cross measurements with a suta give excellent accuracy for diameter measurements, while diametric tapes come second and single measurements with a suta come third in terms of accuracy. When hundreds or thousands of trees in the population are measured, the differences between the measurement methods practically disappear.

There are situations where there is doubt about the position of the diameter measurement, as shown in Figura 7.

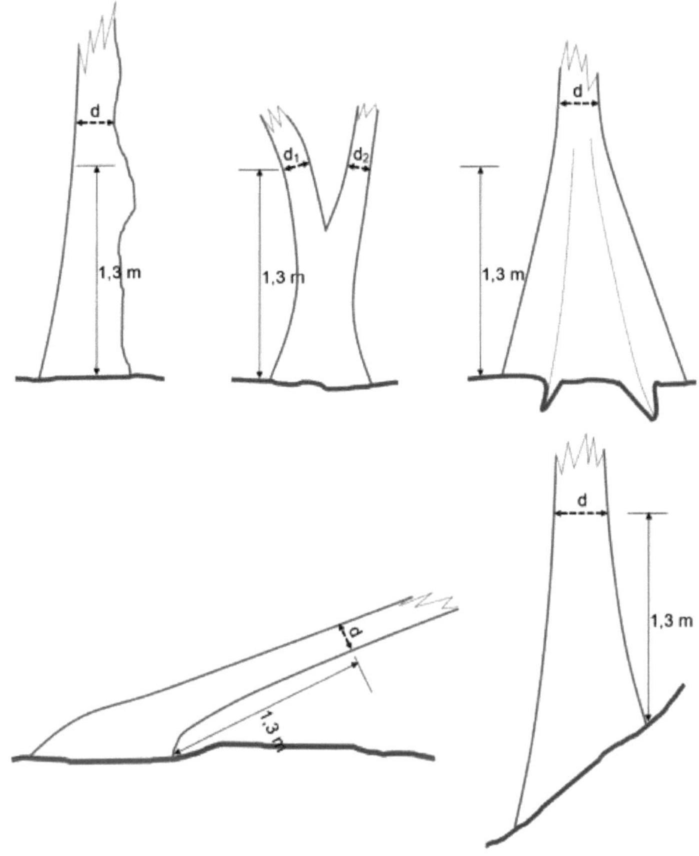

Figura 7 - Diameter (d) measurements in doubtful situations.

When trunk deformations occur at 1.3 m, the diameter should be measured above the deformation. If the tree is forked below 1.3 m, measure all the stems as independent trees. If there are sapopemas at the base of the tree even above 1.3 m, measure above where there is no longer any influence from them. Leaning trees have their diameter measured at 1.3 m from the base below. If the terrain is sloping, the diameter is measured at 1.3 m from the base of the tree at the top of the slope.

Another question that arises is in relation to trees with crooked trunks. In this case, there are two ways of treating the diameter measurement; either it is measured with a special probe that allows measurement inside the crookedness, or this quality is noted for selective processing and the reduction of the transversal surface is estimated by means of images in relation to the transversal surface determined by the diameter measured outside the crookedness; then it is possible to calculate the necessary reduction factor in the basal area (FC_g) of the tree as in Figura 8. The CF_g can be calculated as an average for the species with this type of trunk.

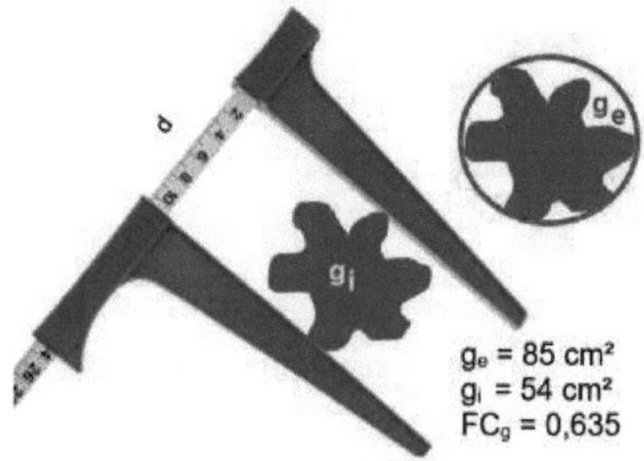

Figura 8 - Basal area correction factor (FC_g) for trees with a torus cross-section, calculated as the internal basal area (g_i) divided by the external basal area (g_e).

Diameter measurements are usually expressed in centimeters to one decimal place, which is the standard for measuring equipment, and averages can be expressed to two decimal places.

There are various expressions for the diameter of tree trunks and stands, as described below.

2.2.2 Diameter at Peel Height (d)$_s$

This is the diameter of the trunk at a height of 1.3 m, excluding the double thickness of bark, calculated by:

$$d_s = d - 2 \cdot e$$

Where: d_s = diameter without bark; d = trunk diameter at 1.3 m height; e = simple bark thickness.

2.2.3 Average basal area tree diameter (d)$_g$

This is the diameter corresponding to the average basal area of the stand, obtained by:

$$d_g = \sqrt{\frac{4\,\bar{g}}{\pi}}$$

Where: dg = diameter corresponding to the average basal area of the stand in cm; $\bar{g}$ = average individual basal area of the stand in cm².

The average individual basal area of the stand is calculated by:

$$\bar{g} = \frac{\sum_{i=1}^{n} g_i}{n} = \frac{G}{N}$$

Where: $\bar{g}$ = average individual basal area of the stand in m²; G = average basal area of the stand per hectare in m²/ha; n = number of trees in the sample; g = average individual basal area in m² of the sampling unit of order i; N = number of trees in the stand per hectare.

2.2.4 Median tree diameter (d)$_M$

It is the diameter corresponding to the tree that divides the population into two of equal frequency, when the diameters are ordered by size. It is the median of the population's diameters.

2.2.5 Median tree diameter of basal area (d)$_{gM}$

It is the diameter corresponding to the tree with the basal area that divides the population into two of equal frequency, when the individual basal areas are ordered by size.

2.2.6 Hohenadl diameters (d-, d+)

These are the diameters of the trees in the Hohenadl sample, calculated by:

$$d_+ = d + s_d$$

$$e$$

$$d_- = d - s_d$$

Where: d = average sample diameter; s_d = standard deviation of sample diameters.

Other Hohenadl diameters used to measure trees at 10%, 30%, 50%, 70% and 90% of their height, from the apex to the base of the tree, are represented by $d_{0,1h}$, $d_{0,3h}$, $d_{0,5h}$, $d_{0,7h}$ and $d_{0,9h}$ respectively.

26

2.2.7 Diameters at different heights (d)$_i$

Diameters at different heights can be represented by d_i, where i is the height considered, for example $d_{5,2}$ represents the diameter at a height of 5.2 meters, or $d_{11,0}$ which represents the diameter at a height of 11.0 meters. Preferably the height should be represented to one decimal place so as not to confuse it with the diameter at a given age. The Finnish suta (Figura 9), when attached to an extendable cable, allows diameters to be measured at different heights and thus allows standing trees to be cubed, provided they are not too tall.

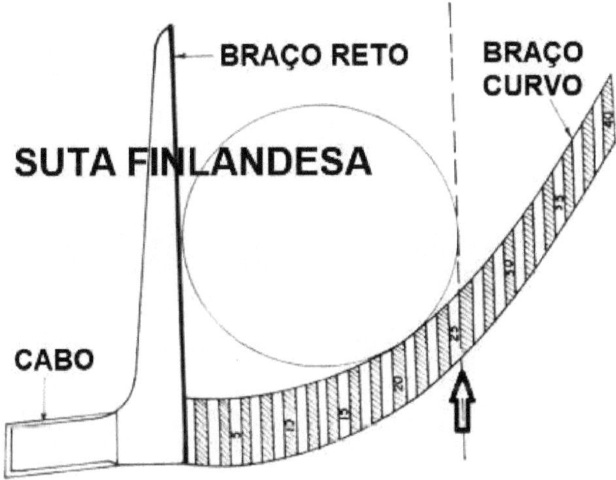

Figura 9 - Finnish suta.

2.2.8 Diameters at different ages (d)$_t$

Diameters at different ages can be represented by d_t, where t is the age considered, for example d_4 represents the diameter of the tree at 4 years of age, or d_{25} which represents the diameter of the tree at 25 years of age. Preferably it should be represented with the age in whole numbers so as not to confuse it with the diameter at a given height described above.

2.2.9 Average diameter ($\bar{d}$)

This is the average diameter of the trees in a sampling unit (plot), calculated using the equation:

$$\bar{d} = \frac{\sum_{i=1}^{n} d_i}{n}$$

Where: $\bar{d}$ = average diameter of the sampling unit; n = number of individuals contained in the sampling unit; d_i = diameter of the tree of order i in the sampling unit; i = number of orders of the trees in the sampling unit.

2.2.10 Weise mean tree diameter (d)$_w$

The average Weise diameter is 40% of the largest trees in the stand, viewed under a frequency distribution. The average Weise tree is close to the tree in the stand with the average volume.

2.2.11 Diameter of the central tree of basal area (d$_z$).

The average diameter of the central basal area tree is obtained by interpolation at the point where the basal area is half the total per hectare. First the diameter class is estimated and then the average diameter is interpolated to 1 mm. The calculation involves classifying the diameters by class and determining the frequency and basal area by class.

2.3 Bark thickness (e)

The bark can be measured using specific instruments or by simply removing a piece of bark in certain positions and measuring it with a caliper.

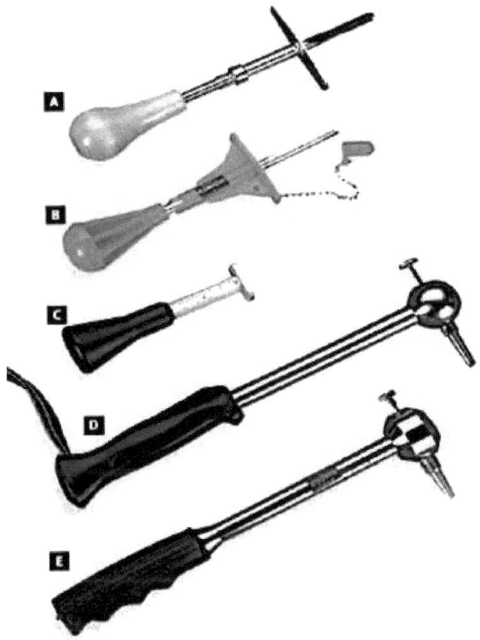

Figura 10 - Bark gauges (A, B, C) and increment hammers (D, E).

Bark thickness is important for obtaining bark-free diameters and bark volumes of trees. It is usually only measured on trees felled for cubing or trunk analysis. The symbol "e" always refers to simple bark thickness.

Many studies have shown that the proportion of bark thickness is maintained along the trunk of the tree and, if it is necessary to determine the bark-free volume without the possibility of felling trees for cubing, it is possible to measure the bark thickness at chest level, calculate the bark-free diameter and determine the proportion of the bark-free diameter to the diameter with bark and, with this proportion, estimate the bark-free diameters along the trunk to obtain the bark-free volumes of each section of the trunk and the tree.

2.4 Basal area

2.4.1 Individual basal area (g)

This is the cross-sectional area of the trunk at a height of 1.3 meters, corresponding to the value of π multiplied by the diameter at a height of 1.3 meters.

It is expressed in square meters, usually and preferably written to five decimal places.

2.4.2 Average individual basal area ($\bar{g}$)

This is the average individual basal area of the trees in a sampling unit (plot), calculated using the equation:

$$\bar{g} = \frac{\sum_{i=1}^{n} g_i}{n}$$

Where: $\bar{g}$ = average individual basal area of the sampling unit; n = number of individuals contained in the sampling unit; g_i = basal area of the tree of order i in the sampling unit; i = number of order of the trees in the sampling unit.

2.4.3 Basal area per hectare (G)

The basal area per hectare is the sum of all the basal areas of the trees in a hectare. It can be calculated using the equation:

$$G = \frac{10000 \sum_{i=1}^{n} g_i}{a}$$

Where: G = basal area in m² per hectare; n = number of trees in the sampling unit; g_i = basal area in m² of the tree of order i in the sampling unit; a = area of the sampling unit in m².

Alternatively:

$$G = \frac{10000 \cdot n \cdot \bar{g}}{a}$$

Where: G = basal area in m² per hectare; n = number of trees in the sampling unit; $\bar{g}$ = average basal area in m² of the trees in the sampling unit; a = area of the sampling unit in m².

2.5 Height

The height of a tree from its base to its top is represented by the letter h, usually displayed in meters to one decimal place, or to a maximum of two decimal places.

Measurement can be carried out using the trigonometric method (Figura 12) or by triangle ratio (Figura 11). In any case, the observer must stand at a distance equivalent to the height of the tree, or up to 1.5 times its height to take the height measurement.

Ideally, the observer should position himself in a place where his eye is level with the base and top of the tree; in practical terms, he should stand at tree level; this makes measuring easier.

2.5.1 Height by ratio of triangles

The triangle ratio method, used by Christen's hypsometer, can be explained with a marker of known height, 5 meters for example, and a transparent ruler of a fixed size of 40 cm. The marker is placed next to the tree trunk and the observer stands back holding the ruler vertically until the entire tree from the base to the top is placed on the ruler from 0 to 40 cm, then the height of the marker next to the tree is measured on the ruler, as shown in Figure 11. Figura 11.

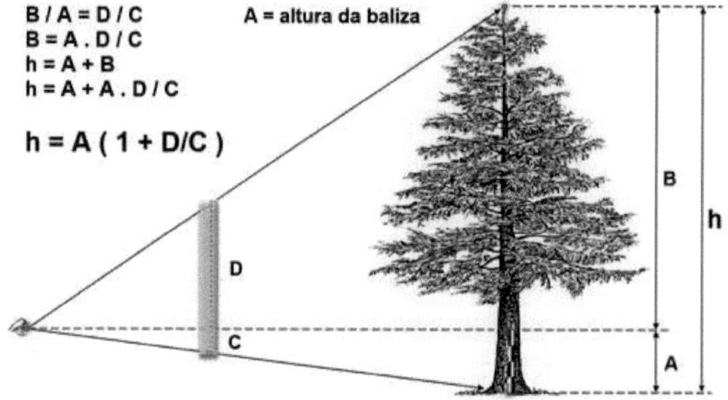

Figura 11 - Measuring height (h) using a ratio of triangles. Image: modified from Bourdo (2001).

The height of the tree will be calculated with reference to the Figura 11considering the relationship

$$B / A = D / C,$$

or

$$B = A \, D / C \qquad (2.4.1.a)$$

Where: A = height of the beacon against the side of the trunk; B = height above the beacon to the top of the tree, this is the unknown part of the height; C = measurement on the ruler corresponding to the height of the beacon; D = height on the ruler corresponding to the distance from the top of the beacon to the top of the tree.

Knowing that the height (h) is the sum of sections A and B, calculated by:

$$h = A + B \qquad (2.4.1.b)$$

Substituting 2.4.1.a into 2.4.1.b, we get that:

$$h = A + A \, . \, D / C,$$

or

$$h = A \, (\, 1 + D / C \,)$$

Where: h = height of the tree; A = height of the marker; C = measurement on the ruler corresponding to the height of the marker; D = height on the ruler corresponding to the distance from the top of the marker to the top of the tree.

In the example given with the 40 cm ruler and the 5 m marker (=A), the measurement of C was 9 cm, so D is 31 cm:

$$h = 5m \, (1 + 31cm / 9cm) = 5m \, (\, 1 + 3.4444 \,)$$
$$h = 5m$$
$$+ \, 5m \, . \, 3.4444$$
$$h = 5m + 17.22m$$
$$h = 22.22 \ m$$

2.5.2 Height by trigonometric method

In the trigonometric method, the observer stands at a point of known distance from the tree to be measured. They measure the angle in relation to the top of the tree and the angle in relation to the base of the trunk. With these three measurements, h_1 and h_2 are calculated as in Figures 11, 12 and 13, and the height (h) is calculated as follows:

> - if the observer's eye is at a level between the base of the trunk and the top of the tree (Figura 12) - add up h_1 and h_2 to find the height of the tree (h);
> - if the observer's eye is at a level below the base of the trunk (Figura 13) - the height (h) is calculated by decreasing h_2 minus h $;_1$
> - if the observer's eye is at a level above the top of the tree (Figura 14) - the height (h) is calculated by decreasing h_1 minus h $._2$

$$h_1 = D \cdot tg\ \alpha$$
$$h_2 = D \cdot tg\ \beta$$

$$h = h_1 + h_2$$

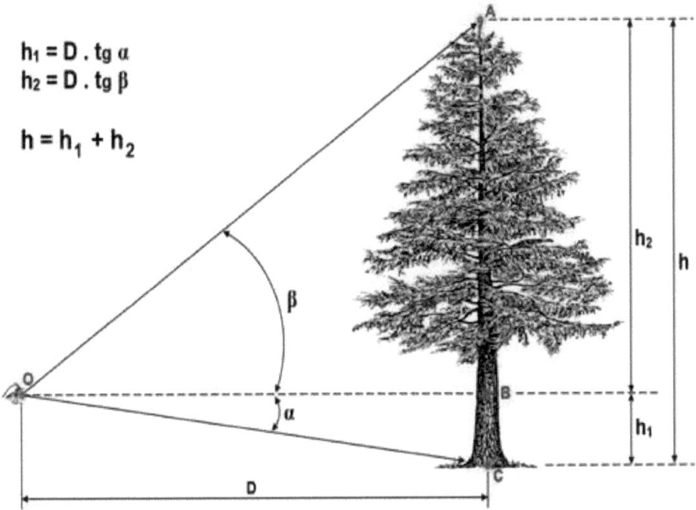

Figura 12 - Observer's eye between the base and top of the tree. Image: modified from Bourdo (2001).

$$h_1 = D \cdot tg\ \alpha$$
$$h_2 = D \cdot tg\ \beta$$

$$h = h_2 - h_1$$

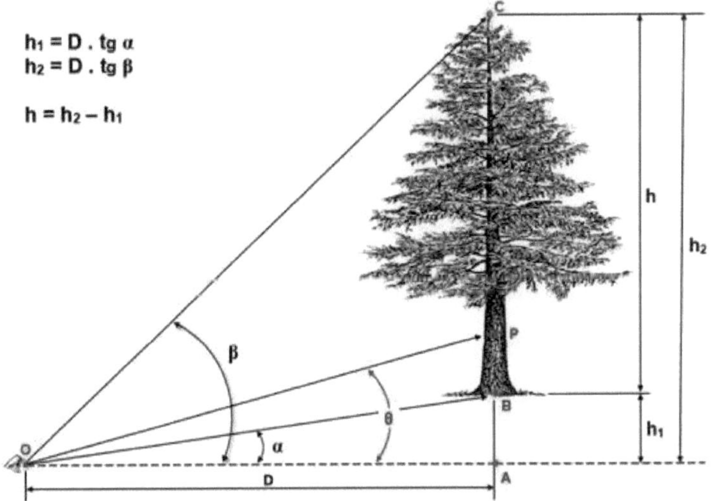

Figura 13 - The observer's eye is below the base of the tree. Image: modified from Bourdo (2001).

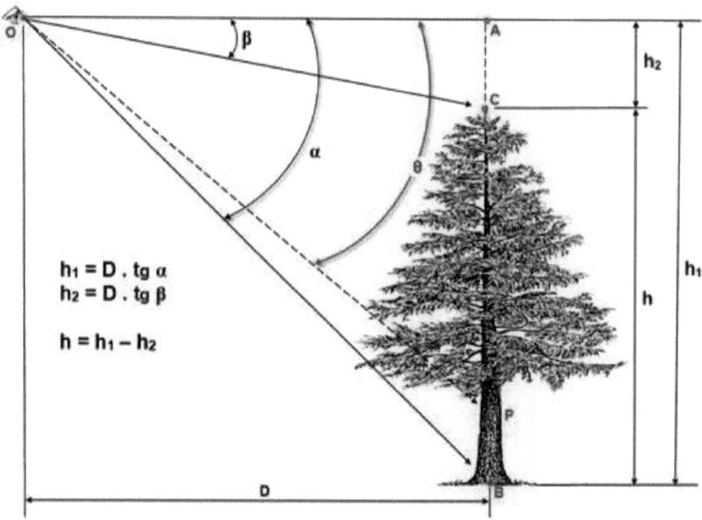

Figura 14 - The observer's eye above the top of the tree. Image: modified from Bourdo (2001).

There is rustic equipment for measuring height, such as the dendrometric clipboard and Christen's hypsometer (Figura 17), but the errors made and the inaccuracy of these devices do not justify their use.

Today, the market offers a variety of precise devices, some of which are affordable, such as the SNDWAY SW-1000A *rangefinder*, which allows precise height measurements to be taken quickly (Figura 15). The device uses a laser beam to take direct measurements of the height of any object with a homogeneous face that reflects the laser, although any reflection of the laser beam from obstacles such as leaves and branches results in an inaccurate measurement. However, it can read precise angles and distances, working as a rangefinder and precision clinometer, making it easy to obtain the precise height of trees by trigonometry.

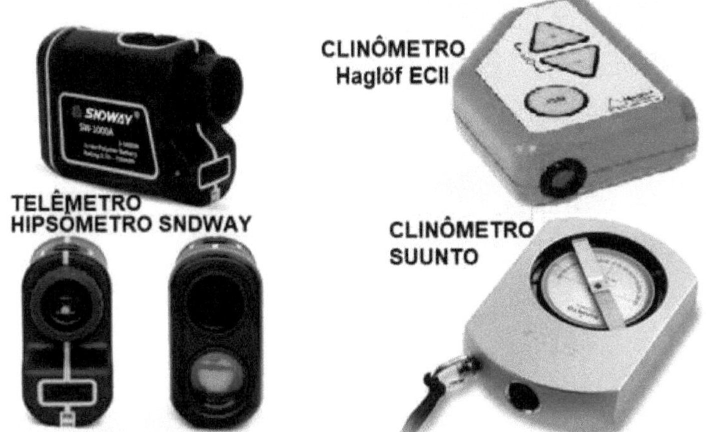

Figura 15 - SNDWAY SW-1000A hypsometer rangefinder, Haglöf ECII clinometer and Suunto clinometer.

Other equipment such as clinometers can measure angles accurately, but require prior knowledge of the distance from the observer to the tree in order to calculate the height. Manufacturer Haglöf offers four clinometer models (CI, HCC, ECII and HCII-R). The ECII offers precise angle and height measurements as long as the distance to the tree Is known. Finnish Suunto provides an analog clinometer for measuring angles, with similar functionality to Haglöf's.

The equipment we have chosen as being the most versatile, allowing precise and very fast direct height measurements, is the Vertex IV (Figura 16), also produced by Haglöf, but its high price has led most people to opt for cheaper equipment. The Vertex IV uses ultrasound technology, making it easy to measure when there is understorey, provided it is not too dense. It is very useful for marking circular plots, as it has a beacon with a transponder holder that can be installed in the center of the plot, helping to demarcate its boundaries. It also lends itself to Bitterlich sampling.

Figura 16 - Vertex IV hypsometer.

Older analog equipment with an accuracy of 0.5 meters, such as the Blume-Leiss hypsometer-telemeter and the Haga hypsometer (Figura 17) allow fast measurements, although less accurate than other laser and ultrasound equipment today.

An intermediately priced hypsometer is Nikon's (Figura 18) which uses laser technology and two measurement modes, with the "two-point height measurement" mode being especially useful for dendrometry.

Among the analog equipment, the most widely used is certainly the Blume-Leiss rangefinder (Figura 19), whose rangefinder allows measurements of distances at 10 m and 15 m and their multiples (20, 30, 40 m), as well as the height of trees, with an accuracy of 0.5 meters, at the selected distance, using the corresponding scale on the instrument.

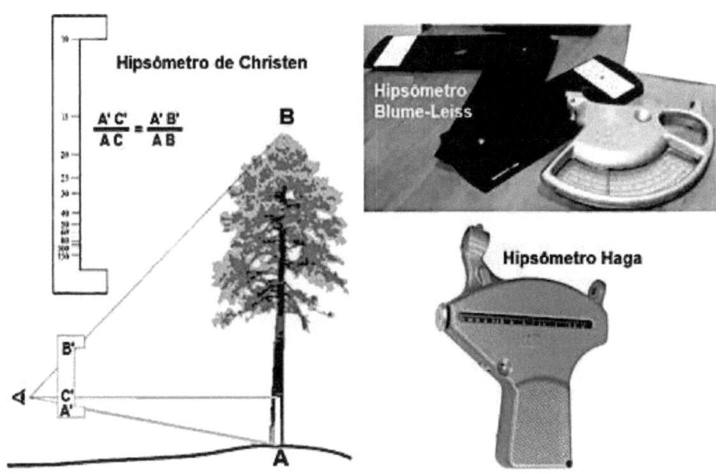

Figura 17 - Christen, Blume-Leiss and Haga hypsometers.

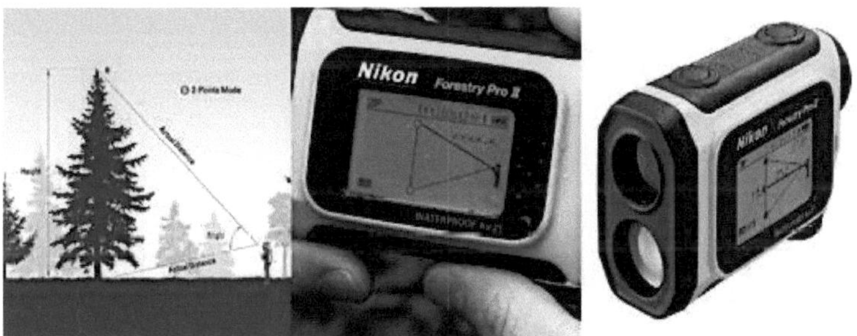

Figura 18 - Nikon Forestry Pro II hypsometer.

Figura 19 - Blume-Leiss hypsometer with rangefinder.

2.5.3 Stem height (h)ₜ

This is the height of the commercial part of the trunk, from the base near the ground to where the trunk can be used, disregarding the branches.

It differs in sympodial and monopodial growth trees.

In trees with sympodial growth, the crown opens at a certain height from where thick branches start and the main trunk ends; the trunk is used up to this point of crown insertion, this is the commercial height, or stem height.

In trees with monopodial growth, the branches are thinner and spread over the entire height and the trunk has a dominant apex. In these trees, commercial height is defined by the minimum diameter of the trunk.

However, it is not always possible to clearly define the height of the stem. In Figura 20you can see some difficulties. The hawthorn (A) has typical sympodial growth and it is easy to establish where the trunk extends and the crown begins. Eucalyptus (B) has a higher branch to the right of the main trunk, which is more vertical, and both could be used for energy, for example; as the stem refers to the main trunk, the commercial height could be defined as the height up to which the diameter of the main trunk allows it to be used. In the case of the araucaria (C), it is a mature specimen and the trunk has good thickness up to almost the top, however the amount of knots near the top of mature araucarias prevents its use, except for energy; in the case of use for sawmills, for example, the height of the stem should be limited to the minimum diameter for sawing, or if the diameter is sufficient even above the point where the branches start, the stem is limited to where the quality of the wood is not impaired by the knots in the branches. Pine (D) typically grows monopodially and, in this case, the height of the stem will be limited by the minimum diameter for commercial use.

Figura 20 - A) Hawthorn; B) Eucalyptus; C) Araucaria; D) Pine.

2.5.4 Lorey height (h)$_L$

Lorey's height is the average height weighted by the basal areas, calculated using the equation:

$$h_L = \frac{\sum_{i=1}^{n} g_i h_i}{\sum_{i=1}^{n} g_i}$$

Where: h_L = Lorey's height; g_i = basal area of tree of order i; h_i = height of tree of order i.

2.5.5 Average height ($\bar{h}$)

It is the arithmetic mean of the heights of the trees in a sampling unit obtained from the equation:

$$\bar{h} = \frac{\sum_{i=1}^{n} h_i}{n}$$

Where: $\bar{h}$ = arithmetic mean height; h_i = height of tree of order i; n = number of trees.

2.5.6 Average basal area tree height (h)$_g$

This is the height corresponding to the tree with the average basal area. It is determined by a regression equation of the type $h_g = f(d_g)$, where d_g is the diameter of the tree with the average basal area ($\bar{g}$).

2.5.7 Average tree diameter height (h)$_d$

This is the height corresponding to the arithmetic mean diameter, obtained from a regression equation of the type $h_d = f(\bar{d})$, where $\bar{d}$ is the arithmetic mean diameter.

2.5.8 Height of tree with median diameter (h)$_{dM}$

This is the height of the tree with the median diameter of the population, obtained from a regression equation of the type h_{dM} = f (d_M) where d_M is the median diameter of the population.

2.5.9 Median basal area tree height (h)$_{gM}$

This is the height of the tree with the basal area corresponding to the median of the population, obtained by a regression equation of the type h_{gM} = f (d_{gM}), where d_{gM} is the diameter of the median tree of basal area (g_M) of the population.

2.5.10 Average Weise height (h)$_w$

Corresponds to the tree with the average Weise diameter (d_w). It is determined by the hypsometric relationship.

2.5.11 Average height of the central basal area tree (h)$_z$

This is the average height of the central basal area tree. It is also obtained from the hypsometric relationship.

2.5.12 Dominant height (h)$_{dom}$

There are many definitions of dominant height, one of the most widely used being Assmann's, because it is easy to obtain, practical and useful. Another definition that should be mentioned is Weise's because it has been used with some frequency.

2.5.12.1 Assmann's dominant height (h)$_{100}$

Assmann's dominant height is defined as: "The height of mean basal area stem from the 100 trees with the largest diameter per hectare (ASSMANN, 1970)", i.e. the height of the mean basal area stem from the 100 trees with the largest diameter per hectare. Currently, a variation of Assmann's dominant height has been used, conceptualized as the "average height of the 100 thickest trees per hectare", because it is easy to obtain, and is represented by h $_{.100}$

The dominant height has been used to construct site index curves because it has the following characteristics:

> - It is little influenced by selective thinning from below and systematic thinning, which are the most commonly applied to forests;
> - It's easy to determine;
> - It can be estimated on top of photos;
> - It has great biological significance, representing the settlement throughout its existence;
> - It is highly correlated with the productivity of forest stands.

However, some care should be taken when it is used to classify sites, as it is influenced by thinning from above, which causes estimates of dominant height values to be lower than the real thing, and it is not representative of inhomogeneous stands, as the oldest trees will have the greatest heights.

2.5.12.2 *Weise dominant height (h)$_0$*

Weise's dominant height is calculated as the average height of 8% of the number of thickest trees, or of the trees that represent 20% of the basal area per hectare of the thickest trees. The trees used vary with the variation in frequency per hectare caused by thinning.

2.5.12.3 *Other definitions of dominant height*

- Average height of the 100 tallest trees per hectare (Hart);
- Average height of 20 % of the tallest trees per hectare;
- Average height of 20 % of the thickest trees per hectare;
- Average height of dominant and codominant trees (Mayer);
- Average height of the 30 tallest trees per hectare (Lewis);
- Height of the tree with a diameter equal to the arithmetic mean of the stand plus three standard deviations (Näslund);
- Height corresponding to the average diameter of 20% of the thickest trees in the stand (Weise).

2.5.13 Difficulties in measuring height

2.5.13.1 Leaning trees

It is necessary to measure the height from the ground to the apex of the tree, which will be the right side of a right triangle, and the angle (α) of inclination of the tree (Figura 21).

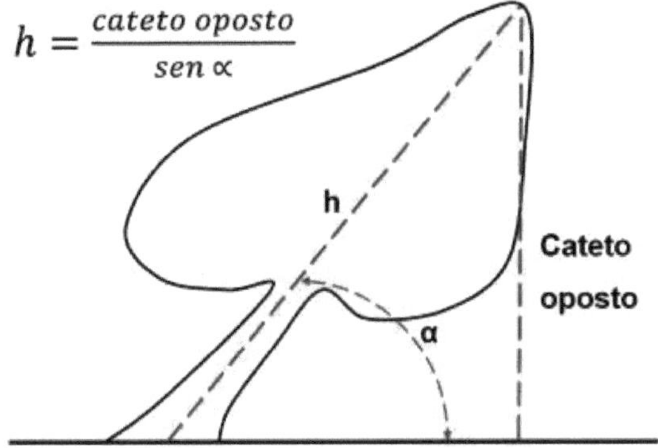

Figura 21 - Measuring the height of leaning trees

The height (h) is the hypotenuse of the triangle and will be calculated by:

$$h = \frac{cateto\ oposto}{sen\ \alpha}$$

Where: h = height of the tree; opposite cateto = height from the ground to the apex of the tree.

2.5.13.2 Trees with irregular tops

Sometimes you can't define exactly where the top of the tree is. In these cases, you have to follow the main trunk and try to define where the top is (Figura 22). There is no special equation for this.

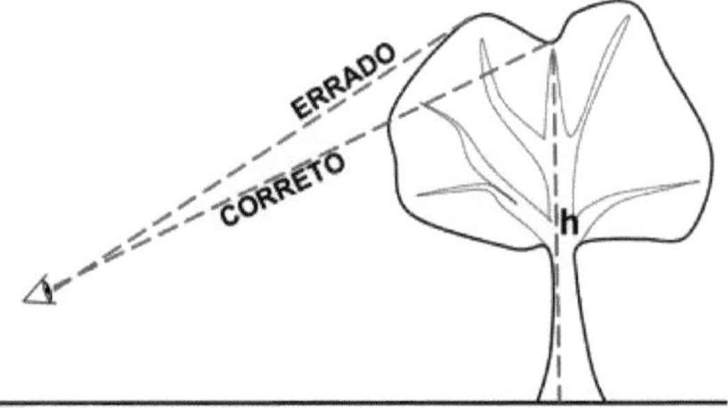

Figura 22 - Measuring the height of trees with irregular tops.

2.6 Shape quotients

Shape quotient is the ratio between two diameters at different heights on the trunk.

2.6.1 Artificial shape quotient (k)

The artificial shape quotient (k) is when the trunk diameter at half height is divided by the diameter at 1.3 m height:

$$k = d_{0,5h} / d$$

Where: k = artificial shape quotient; $d_{0,5h}$ = diameter at 50% of tree height; d = diameter at 1.3 m height.

2.6.2 Absolute shape quotient (k)ₐ

It is calculated as the ratio between the diameter taken at half height plus 1.3 m and the diameter at breast height (d):

$$k_a = d_{[0,5.(h+1,3)]} / d$$

Where: ka = absolute shape quotient; $d_{[0,5.(h+1,3)]}$ = diameter taken at half the height plus 1.3 m; d = diameter at 1.3 m height.

2.6.3 Other shape quotients

Other shape quotients are indicated by subscripts, for example:

- $k_{5,3/1,3} = d_{5,3} / d_{1,3}$, or simply $k_{5,3}$, since $d_{1,3}$ can be represented by d without the subscript 1.3;
- $k_{0,5h/0,1h} = d_{0,5h} / d_{0,1h}$

2.7 Volume

2.7.1 Individual volume (v)

The individual volume of the tree (v) and is always with bark, unless otherwise stated. It is the total volume of the trunk from the base to the top of the trunk, excluding branches. The individual volume is expressed in cubic meters (m^3) and represented to 4 decimal places.

Measurements and determination of the strict volume of trees are generally carried out on felled trees using established deterministic equations such as those of Hohenadl, Hossfeld, Simony, Newton, Huber and Smalian. Figura 32.

The volume of trees, except for the Hohenadl method, is determined separately for the volume of the stump and the upper tip. The volume of the stump is calculated as a cylinder with the length (L_0) of the stump and the cross-sectional area at the top (g_0) of the stump. Calculating the volume of the upper tip of the trunk in sympodial trees does not differ from the other sections. However, in monopodial trees, the volume of the upper tip (v_n)

of the trunk is calculated as if it were a perfect cone, i.e. 1/3 of the volume of a cylinder with the diameter of the base of the tip (d_n) and the length of the tip (L_n).

2.7.2 Individual volume without shell (v)$_s$

The individual volume without bark (v_s) is determined in the same way as the volume with bark, but using the diameters without bark.

2.7.3 Average individual volume ($\bar{v}$)

This is the arithmetic mean volume of the trees in a sampling unit (plot), calculated using the equation:

$$\bar{v} = \frac{\sum_{i=1}^{n} v_i}{n}$$

Where: $\bar{v}$ = average individual volume of the sampling unit; n = number of individuals contained in the sampling unit; v_i = volume of the tree of order i in the sampling unit; i = order number of the trees in the sampling unit.

2.7.4 Volume per hectare (V)

The volume per hectare is the sum of all the basal areas of the trees in a hectare. It can be calculated using the equation:

$$V = \frac{10000 \sum_{i=1}^{n} v_i}{a}$$

Where: V = volume in m³ per hectare; n = number of trees in the sampling unit; v_i = volume of tree of order i in the sampling unit; a = area of the sampling unit in m².

Alternatively:

$$V = \frac{10000 \cdot n \cdot \bar{v}}{a}$$

Where: V = volume in m³ per hectare; n = number of trees in the sampling unit; $\bar{v}$ = average volume in m³ of the trees in the sampling unit; a = area of the sampling unit in m².

2.8 Form factor (f)

The artificial form factor (f) or ($f_{1,3}$) is the ratio between the individual volume of the tree (v) and its height (h) multiplied by the basal area (g) at 1.3 m height. The form factor is a variable with no unit of measurement. The shape of trees varies with age, sociological position in the vertical strata, site quality and other factors, and is rarely used because of this. This variable allows the volume of a tree to be approximated using the equation:

$$v = f\,h\,g$$

Where: v = individual volume; f = artificial form factor; h = tree height; g = basal area of the tree at 1.3 m height.

The natural form factor ($f_{0,1h}$) is the ratio between the individual volume of the tree (v) and its height (h) multiplied by the basal area (g) at 10% of the height from the base of the tree, obtained from the equation:

$$f_{0,1h} = v / h\; g_{0,1h}$$

Where: $f_{0,1h}$ = natural form factor; h = tree height; $g_{0,1h}$ = basal area at 10% of height from base.

There are other form factor expressions, but due to the inaccurate results of all of them, including the two already described, they are not very useful in Forest Engineering and will not be dealt with in this work.

2.9 Increments (i, I and P)

The symbols i and I represent the increase in tree size over the course of a year, unless otherwise stated. The lower case letter i is used for the increment of individual trees and the upper case letter I is used for the increment per hectare. This increment is also referred to as the "current annual increment", meaning that the tree has increased in size from one year to the next.

Individual increase averages should be graphed with the letter $\bar{\imath}$ superimposed by a bar, meaning the arithmetic mean of the increase in one year.

Increases over periods of more than one year should be subscripted by the age at which the period begins and ends; for example: $i_{15\text{-}25}$ represents the increase between the ages of 15 and 25. The average annual increment over a period should be represented with the symbol superimposed by a slash; example: $\bar{\imath}_{15-25}$ is the average increment of trees from 15 to 25 years of age.

Increases are always represented with the letter i followed by the symbol of the variable to which the increase refers, such as i_d , $\bar{\imath}_v$, I_G , I_V representing the individual increment of a tree in diameter, the average individual increment of trees in volume, the increment in basal area per hectare and the increment in volume per hectare, respectively.

P represents the percentage increase of a given variable in one year. For example: if the diameter of a particular tree was measured at 17 and 18 years of age, at 31 and 33 cm respectively, the diametric increase was 2 cm, and the tree had a percentage increase of P = 100 (33-31) / 31 = 6.45% from 17 to 18 years of age.

When considering a growth equation, the current annual increase can be determined by means of the first derivative of the integral growth equation and the proportion of the current annual increase by its second derivative.

2.10 Canopy dimensions

Photosynthesis takes place in the treetops, which promotes growth. Studying the crowns and their proportions helps determine the density of stands, competition between trees and planning management actions to obtain products with the desired dimensions and maximum productivity. The main dimensions of the tree crown are shown in Figura 24.

2.10.1 Radius and crown diameter

Four crown radii are usually measured in the north (RN), east (RL), south (RS) and west (RO) directions. The crown diameter (CD) is calculated as twice the average of the four radii, or half the sum of the 4 radii, using the equation:

$$DC = (RN+RL+RS+RO)/2$$

Where: DC = crown diameter; RN, RL, RS, RO = crown radii, North, East, South and West, respectively.

The measurement of crown radii can be aided by a densitometer (Figura 23), which makes the measurements more accurate.

Figura 23 - Densitometers: auxiliary instruments for measuring crown radii.

It is also possible to measure the crown radii using a bubble level. The observer should stand perpendicular to the crown radii and align the marker vertically towards the tip of the branch to measure the length of the radius (Figura 24).

Figura 24 - Measurement of crown radii using a bubble level. Tree crown dimensions: h = tree height; h_{bc} = crown base height; CC = crown length; RN, RL, RS, RO = crown radii, North, East, South and West, respectively; SPC = crown projection surface.

2.10.2 Crown base height (h)$_{bc}$

The height of the base of the canopy is not always easy to determine, and it is necessary to take into account the green branches that actually contribute to growth. Sometimes there are small branches in the lower part that have minimal contribution to growth and should be disregarded. Also, the insertion point of the branches should not be taken into account. What is important is the foliage, which is the part of the crown that contributes to photosynthesis. The height of the base of the crown can be conceptualized as the height where the foliage that contributes to the tree's growth begins (Figura 24).

2.10.3 Cup length (CC)

Crown length (CL) is calculated as the difference between tree height (h) and crown base height (hbc).

2.10.4 Canopy projection surface (SPC)

The crown projection surface (SPC) is calculated as a function of the crown diameter (DC) using the equation:

$$SPC = \pi \frac{DC^2}{4}$$

Where: SPC = crown projection surface; DC = crown diameter.

3 FORESTRY STATISTICS

Statistics is the science of collecting, organizing, summarizing, analyzing, interpreting and presenting population or sample data.

3.1 Univariate analysis

Univariate statistics is the part of statistics that deals with describing and making inferences about a single variable in a population or sample.

3.1.1 Descriptive statistical analysis

This is the part of statistics that aims to summarize or describe the distribution of a single variable.

The techniques and calculations used in descriptive statistics are mainly:

- ➤ Measures of central tendency;
- ➤ Measures of dispersion;
- ➤ Probability distributions;
- ➤ Histograms.

3.1.1.1 Measures of central tendency or position

3.1.1.1.1   Average

It is the ratio between the sum of the values of observations on a given variable (x) and the number of observations made (n), calculated using the formula:

$$\bar{x} = \frac{\sum_{i=1}^{n} x_i}{n}$$

Where: $\bar{x}$ = average of the observations on variable x; x_i = value of the observation of variable x of order i; n = number of observations made.

3.1.1.1.2 Median

The median is the value of a variable x that divides the population into two equal parts. Half of the values of the observations of x are smaller than the median and the other half are larger.

3.1.1.1.3 Fashion

This is the most frequent value of a variable x, i.e. the value of x that occurs most often in the population.

3.1.1.2 Measures of dispersion

3.1.1.2.1 Sample variance

It is the average of the squares of the differences between the observations of a variable x and its mean. It is calculated using the formula:

$$s^2 = \frac{\sum_{i=1}^{n} x_i - \bar{x}}{n - 1}$$

Where: s^2 = sample variance; x_i = value of x observation of order i; $\bar{x}$ = average of x observations.

It can also be calculated using the formula:

$$s^2 = \frac{\sum_{i=1}^{n} x_i^2 - \frac{\left(\sum_{i=1}^{n} x_i\right)^2}{n}}{n - 1}$$

3.1.1.2.2   Standard Deviation

It is the square root of the variance of a variable x, calculated by the formula:

$$s = \sqrt{s^2}$$

3.1.1.2.3   Coefficient of Variation

The coefficient of variation is the percentage that the standard deviation represents over the mean, calculated using the formula:

$$CV = 100 \cdot s / \bar{x}$$

Where: CV = coefficient of variation in percentage; s = standard deviation; $\bar{x}$ = average of x observations.

3.1.1.3 Probability distributions

Probability distribution is the proportion of occurrence of a phenomenon in relation to the whole, dependent on chance, modeled mathematically.

Probability distribution describes the random behavior of a phenomenon dependent on chance.

A probability distribution function is a mathematical model that relates a certain value of the variable under study to its probability of occurrence.

There are two types of probability distribution:
> Continuous Distributions - When the variable being measured is expressed on a continuous scale, as in the case of a dimensional characteristic; example: tree diameter.

> Discrete Distributions - When the variable being measured can only take on certain values, such as integer values: 0, 1, 2, etc.; example: live tree=0, dead tree=1.

3.1.1.3.1   Continuous

3.1.1.3.1.1 Normal distribution

The probability density function of the Normal distribution of a variable x, with mean µ and variance σ², is defined as:

$$f(x) = \frac{1}{\sigma\sqrt{2\pi}} \exp\left(-\frac{(x-\mu)^2}{2\sigma^2}\right)$$

Where: µ = mean, or location parameter; σ = standard deviation, or scale parameter; [illegible]x[illegible].

The cumulative distribution function is expressed by:

$$F(x) = \frac{1}{2}\left(1 + \operatorname{erf}\frac{x-\mu}{\sigma\sqrt{2}}\right)$$

$$\operatorname{erf}(x) = 2\Phi\left(x\sqrt{2}\right) - 1$$

$$\Phi(x) = \frac{1}{\sqrt{2\pi}} \int_{-\infty}^{x} e^{-\frac{1}{2}t^2}\, dt$$

An example of the fit of the normal distribution of diameters is shown in Figura 25.

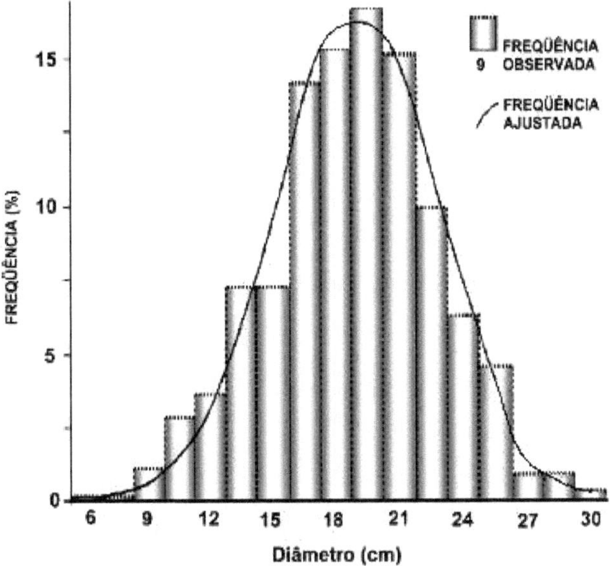

Figura 25 - Normality of the frequencies by diameter class, of the trees contained in the useful area of the plots, at 15 years of age.

3.1.1.3.1.2 Lognormal distribution

A random variable x has the Lognormal distribution when its logarithm $y-Ln(x)$ has the normal distribution with mean $e^{\mu+\sigma^2/2}$, variance $(e^{\mu^2}-1)\,e^{2\mu+\sigma^2}$ and probability density function given by:

$$f(x;\mu,\sigma) = \frac{1}{x\sigma\sqrt{2\pi}} \exp\left[-\frac{(\ln(x)-\mu)^2}{2\sigma^2}\right]$$

Where: for $x > 0$, where μ and σ are the mean and standard deviation of the logarithm of the variable (by definition, the logarithm of the variable is normally distributed).

3.1.1.3.1.3 Weibull distribution

Ernst Hjalmar Waloddi Weibull (Kristianstad County, June 18, 1887 - Annecy, October 12, 1979) was a Swedish engineer and mathematician.

The Weibull distribution is very flexible and can take a variety of forms.

It has been used to model diameter distributions in planted forests.

The formulas developed by Weibull for probability distributions are as follows:

➤ Cumulative distribution function with two parameters

$$F(x; \alpha, \beta) = 1 - e^{-(x/\beta)^\alpha}$$

➤ Probability density function with two parameters

$$f(x; \alpha, \beta) = \frac{\alpha}{\beta^\alpha} \, x^{\alpha-1} \, e^{-(x/\beta)^\alpha}$$

➤ Cumulative distribution function with three parameters

$$F(x) = 1 - \exp\left(-\left(\frac{x - \alpha}{\beta}\right)^\gamma\right)$$

➤ Probability density function with three parameters

$$f(x) = \left(\frac{\gamma}{\beta}\right) \cdot \left(\frac{x - \alpha}{\beta}\right)^{\gamma-1} \cdot \exp\left(-\left(\frac{x - \alpha}{\beta}\right)^\gamma\right)$$

3.1.1.3.1.4 Beta distribution

$$f_x(x) = \frac{1}{\beta(a, b)} \cdot x^{\alpha-1} \cdot (1 - x)^{\beta-1}$$

The probability density function of the Beta distribution satisfies the differential:

$$f'(x) = f(x)\,\frac{(\alpha+\beta-2)x-(\alpha-1)}{(x-1)x}$$

The cumulative probability distribution is calculated by:

$$F(x;\alpha,\beta) = \frac{B(x;\alpha,\beta)}{B(\alpha,\beta)}$$

$$= \frac{\int_0^x t^{a-1}(1-t)^{b-1}\,dt}{B(\alpha,\beta)}$$

$$= I_x(\alpha,\beta)$$

Where: where $B(x;\alpha,\beta)$ is the incomplete beta function and $I_x(\alpha,\beta)$ is the regularized incomplete beta function.

3.1.1.3.1.5 Gama Distribution

A variable x with Gamma distribution has mean $k\theta$, variance $k\theta^2$ and probability density function defined by:

$$f(x;\alpha,\beta) = \frac{\beta^\alpha x^{\alpha-1}e^{-\beta x}}{\Gamma(\alpha)} \quad \text{for } x > 0 \text{ and } \alpha,\beta > 0$$

The cumulative probability distribution is expressed by:

$$F(x; \alpha, \beta) = \int_0^x f(u; \alpha, \beta)\, du = \frac{\gamma(\alpha, \beta x)}{\Gamma(\alpha)}$$

3.1.1.3.1.6 Exponential distribution

In the Exponential distribution, the random variable is defined as the interval between two occurrences (time, space, etc.). The probability density function is given by:

$$f(x) = \frac{1}{\mu} e^{-x/\mu}, \quad x \geq 0$$

Where: f(x) = probability of x occurring; μ = average interval between occurrences; x = value of the variable.

The Exponential distribution model can be expressed as:

$$f(x) = \lambda e^{-\lambda x}; \qquad t \geq 0$$

Where: λ > 0 and is a constant parameter = 1 / μ.

The mean and standard deviation of the exponential distribution are calculated using:

$$\mu = \frac{1}{\lambda} \qquad \sigma = \frac{1}{\lambda}$$

Where: μ = mean; σ = standard deviation.

Example: since the average number of people served at the bank teller is λ= 6/min, the average time between services is 1/ λ = 1/6 min = 10 sec.

Natural forests tend to have a negative exponential diametric distribution, as shown in Figura 26. The cumulative distribution model can be expressed by:

$$N_i = b_0 \cdot e^{-b_1 \cdot d_i}$$

Where: N_i = number of trees per hectare accumulated up to diameter class of order i; d_i = upper limit of diameter class of order i; b_0 , b_1 = model parameters.

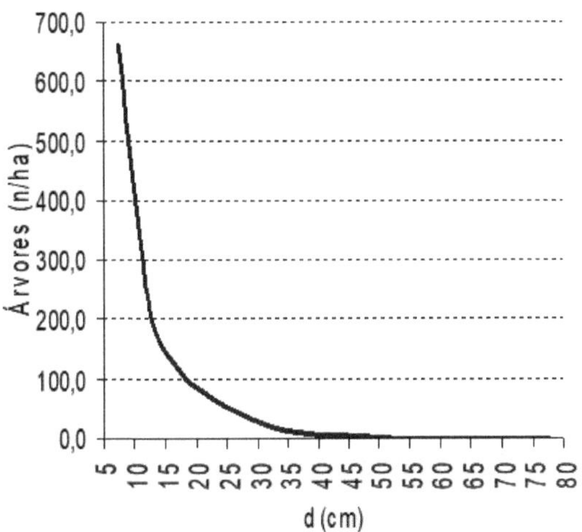

Figura 26 - Cumulative diametric distribution of a natural forest.

3.1.1.3.2   Discrete Distributions

3.1.1.3.2.1 Binomial distribution

There are no position or scale parameters in the binomial distribution.

It is suitable for describing the results of a random variable in just two mutually exclusive classes, or categories, such as:

$$P \ (success) + P \ (failure) = 1$$

The probability density function of the binomial distribution is:

$$b(x; n, p) = \binom{n}{x} p^x (1 - p)^{n-x}$$

where: $\binom{n}{x}$ = combination of n values taken from x into x, in MS-Excel COMBIN(n,x); p = is a constant, variable or numerical expression that specifies the probability of success ($0 \leq p \leq 1$); n = number of independent trials; x = number of successful trials.

In MS-Excel, the function is as follows:

=DISTR.BINOM(Num_s; Attempts; Probability_s; Cumulative)

Where: Num_s = x = successes; Trials = n = total number of trials; Probability_s = p (probability of success in each trial); Cumulative = FALSE for probability density, TRUE for cumulative probability.

The cumulative binomial distribution is:

$$B(x; n, p) = \sum_{y=0}^{x} b(y; n, p)$$

Distribution application conditions:

➢ *n* repetitions of the experiment are made, where n is a constant;

➢ there are only two possible outcomes in each repetition, called success and failure;

➢ the probability of success *(p)* and failure *(1- p)* remain constant for all repetitions;

➢ the repetitions are independent, i.e. the result of one repetition is not influenced by other results.

Binomial distributions with p=0.5 are symmetrical. They are asymmetrical when p=0.5. Asymmetry increases as p approaches 0 (positive asymmetry) or 1 (negative asymmetry).

3.1.1.3.2.2 Bernoulli distribution

The simplest experiments in which we observe the presence or absence of some characteristic are known as Bernoulli tests. Some examples:

> ➢ Toss a coin and see if it comes up heads or tails;
> ➢ Roll a dice and see if there are sixes or not;
> ➢ Check whether a tree in the forest is attacked by leaf-cutting ants or not.

The Bernoulli distribution is a special case of the Binomial Distribution, which models the number of successes in a series of binomial trials. A Bernoulli Variable is any random variable whose only possible values are 0 or 1.

This distribution is specified with a single parameter (p), as follows:

$$f(k;p) = p^k(1-p)^{1-k} \quad \text{para } k \in \{0,1\}.$$

$$f(k;p) = \begin{cases} p & \text{se } k = 1, \\ 1-p & \text{se } k = 0. \end{cases}$$

Where: f(k,p) = probability of success or failure; k = proportion of the event occurring (0 or 1); p = proportion of the event considered occurring successfully.

The two possible events in each case are called success (1) and failure (0). The Bernoulli trial is characterized by a random variable X, defined by X=1, if success; X=0, if failure.

An experiment of size N can be interpreted as tossing a coin N times and counting the number of heads and tails of the result. If p = 0.5, there is an equal probability of success or failure. In the case of the coin toss, it is considered reliable. If p ≠ 0.5, the coin is flawed.

3.1.1.3.2.3 Poisson distribution

The Poisson distribution is suitable for describing situations where there is a probability of occurrence in a continuous field or interval, usually time or area.

Application conditions:
- ➢ The number of occurrences during any interval depends solely on the length of the interval;
- ➢ Occurrences occur independently, i.e. an excess or lack of occurrences in one interval has no effect on the number of occurrences in another interval;
- ➢ The possibility of two or more occurrences happening in a short interval is very small compared to that of a single occurrence.

The Poisson distribution is characterized by a single parameter λ which represents the average rate of occurrence per unit of measurement.

The equation to calculate the probability density of x occurrences is given by:

$$p(x) = \frac{e^{-\lambda} \lambda^{x}}{x!}, \qquad x = 0, 1, ..., n$$

The mean and variance of the Poisson distribution are:

$$\mu = \lambda$$

$$\sigma^2 = \lambda^2$$

A typical application of the Poisson distribution is in quality control for the number of defects (non-conformities) that occur per unit of product (per m², per volume or per time, etc.).

The Poisson cumulative function is calculated by:

$$P(x) = \sum_{k=0}^{x} \frac{e^{-\lambda} \lambda^{x}}{k!}$$

3.1.2 Inferential statistical analysis

It is the part of univariate statistics that aims to analyze the data from a sample, obtaining results and, from them, making inferences for the entire population, i.e. it aims to generalize the results of the sample to the population studied.

Inferential statistics uses techniques based on probability theory; it is the basis of sampling theory, which is used in forest inventories.

3.2 Bivariate statistical analysis

It is the part of statistics that deals with the study of relationships between pairs of variables. The correlation between two variables in a sample can be studied using correlation indices, among the most commonly used are the following:

- ➢ Pearson's Correlation Coefficient - measures the correlation between two continuous variables;
- ➢ Spearman's rank correlation coefficient - measures the correlation between two discrete variables;
- ➢ Covariance - a measure of the degree of numerical interdependence between two random variables.

The relationships between two variables can be summarized using graphs and tables.

3.2.1 Quantitative measures of dependency

3.2.1.1 Pearson's correlation coefficient

Pearson's simple sample correlation coefficient is determined by the equation:

$$r_{xy} = \frac{\sum_i ((x_i - \bar{x})(y_i - \bar{y}))}{\sqrt{\sum_i (x_i - \bar{x})^2 \sum_i (y_i - \bar{y})^2}}$$

Where: r_{xy} = Pearson's correlation coefficient; x_i = the value of variable x of order i; y_i = the value of variable y of order i; $\bar{x}$ = sample mean of variable x; $\bar{y}$ = sample average of the y variable.

Pearson's weighted sample correlation coefficient is calculated using the equation:

$$r_{xy} = \frac{\sum_i w_i (x_i - \bar{x}_w)(y_i - \bar{y}_w)}{\sqrt{\sum_i w_i (x_i - \bar{x}_w)^2 \sum_i w_i (y_i - \bar{y}_w)^2}}$$

Where: r_{xy} = Pearson's correlation coefficient; x_i = the value of variable x of order i; y_i = the value of variable y of order i; $\bar{x}_w$ = weighted sample mean of variable x; $\bar{y}_w$ = weighted sample mean of the y variable; w_i = the weight of the observation of order i.

The probability values for Pearson's correlation coefficient are computed using the equation:

$$t = (n-2)^{1/2} \left(\frac{r^2}{1-r^2} \right)^{1/2}$$

According to the t-distribution with n-2 degrees of freedom, where r is the Pearson correlation of the sample.

3.2.1.2 *Spearman's rank correlation coefficient*

Spearman's rank order correlation coefficient (ranks) is a non-parametric measure of association based on the rankings of data values, calculated using the following formula:

$$\theta = \frac{\sum_i ((R_i - \bar{R})(S_i - \bar{S}))}{\sqrt{\sum_i (R_i - \bar{R})^2 \sum (S_i - \bar{S})^2}}$$

Where: θ = Spearman's correlation coefficient; R_i = rank of x_i ; S_i = rank of y_i ; $\bar{R}$ = average of R values$_i$; $\bar{S}$ = average of the values of S .$_i$

The probability values for Pearson's correlation coefficient are computed using the equation:

$$t = (n-2)^{1/2} \left(\frac{r^2}{1-r^2} \right)^{1/2}$$

Where: t = t distribution with n-2 degrees of freedom; r = Spearman's correlation coefficient of the sample.

3.2.1.3 Covariance

Covariance or joint variance is a measure of the degree of interdependence or linear numerical interrelationship between two random variables X and Y. The covariance is calculated using the following formula:

$$Cov(xy) = S_{xy} = \frac{\sum_{i=1}^{n} x_i \cdot y_i - \frac{\sum_{i=1}^{n} x_i \cdot \sum_{i=1}^{n} y_i}{n}}{n - 1}$$

Or:

$$S_{xy} = \frac{\sum_{i=1}^{n}(x_i - \bar{x}) \cdot (y_i - \bar{y})}{n - 1}$$

Where: $Cov(xy) = S_{xy}$ = covariance between variables x and y; x_i = value of variable x of order i; y_i = value of variable y of order i; $\bar{x}$ = average of the values of x_i ; $\bar{y}$ = average of y values$_i$; n = number of observations.

Analysis of covariance is used to verify the need to use independent functions for the covariates, for example, to describe the hypsometric relationship and volume, where the covariate can be represented by the site index or age. According to Storck and Lopes (1998), when calculating covariance it is possible to check the assumption of homogeneity of the estimated linear regression coefficients ($\hat{\beta}$) by calculating a coefficient for each class separately and then testing their equality.

Schneider (1998) uses the "F" value of the interaction of the covariate with the estimated dependent variable ($\hat{Y}$), calculated by the sum of squares of corrected products, to compare the differences between slopes and uses the "F" value calculated for the covariate through the sum of squares of corrected products of the model without interaction to compare the difference between levels, when there is no difference between slopes. In SAS, the sums of squares of corrected products are obtained using option SS3 of the MODEL statement in the GLM procedure.

4 REGRESSION ANALYSIS

Regression analysis is a statistical tool used in forestry to adjust and evaluate stochastic models, with the aim of estimating variables that are difficult and time-consuming to measure, as a function of variables that are measured more easily and quickly.

Equations can be adjusted using statistical software and spreadsheets based on the simple and generalized least squares methods.

The quality of regression models is directly influenced by the quality of the sample. An adequate sample, with a sufficient number of individuals for the analysis of regression variance and allowing for verification and validation, is essential for assessing their quality.

In the past, researchers indiscriminately transformed the variables in the hope of finding better models, but this practice makes regression analysis unfeasible, as the logarithmic discrepancy distorts the mean square values of the errors and, consequently, Snedecor's F in the regression variance analysis, and it is impossible to estimate the true statistics by approximations, as Meyer and Furnival tried to do. The calculated values are like closed packets and can only be obtained by extracting the logarithm of the dependent variable and redoing the analysis of variance of the regression with the observed and estimated values without the logarithmic transformation.

The terms "model verification" and "model validation" are commonly used to indicate model evaluation. Model validation is one of the phases of model evaluation, involving the process of determining whether the performance of a model is of an acceptable level for its purpose (BURKHART and TOMÉ, 2012).

The simplest evaluation of a model should involve analyzing the variance of the regression, calculating the Coefficient of Determination (R^2) and the Standard Error of Estimate as a percentage of the mean of the observations (Syx%), also known as the Coefficient of Variation, as well as validating it by analyzing the residuals.

Regression models are validated by checking three assumptions regarding the analysis of residuals: normality of the residuals, homoscedasticity of the variance of the residuals and independence of the residuals. Some authors consider that graphical analysis of residuals may be sufficient for validation, however, there are no universally accepted rules for graphical analysis. Floriano (2004) developed a model for graphical analysis that is useful for systematizing residue analysis. However, the use of statistical tests is recommended.

Other complementary tests are used to evaluate models, but are not considered essential, such as: checking for *lack of fit*, checking for *bias*, evaluating *model* efficiency, correlation over time, correlation between components and the use of an independent validation sample.

4.1 Regression analysis of variance

A Tabela 1 shows the analysis of variance of the regression of a function $y=f(x_1, x_2, ..., x_n)$.

Fisher's F value, found in the analysis of variance of the regression, gives the value to check whether the regression model under analysis is significant. Statistical software usually reports the probability of significance of F (Pr>F), so there is no need to use F tables. When Pr>F is less than or equal to 0.01, F is "highly significant", or significant at the 1% probability level; when Pr>F is greater than 0.01 and less than or equal to 0.05, F is said to be "significant", being significant at the 5% probability level; and when Pr>F is greater than 0.05, the regression model is "not significant". The significance level can be represented by placing the following symbols in superscript next to the F value:

> ➤ ** - for highly significant F;
> ➤ * - for significant F;
> ➤ ns - for non-significant F.

TABELA 1 - Regression analysis of variance .

Variation Factor (FV)	Degrees of Freedom (GL)	Sum of Squares (SQ)	Mean Square (QM)	F value
Regression	p-1	$\Sigma(ye - ym)^2$	SQreg/GLreg	QMreg/QMres
Waste	n-p	$\Sigma(y - ye)^2$	SQres/GLres	
Total	n-1	$\Sigma(y - ym)^2$		

Where: y = observed value of the dependent variable; ye = estimated value of the dependent variable; ym = mean of the observations of y; GLreg = degrees of freedom of the regression; SQres = degrees of freedom of the residual; SQreg = sum of squares of the regression; SQres = sum of squares of the residuals; QMreg = mean square of the regression; QMres =

mean square of the residuals = variance. Source: Bussab (1986).

4.1.1 Standard Error of Estimates (S)$_{yx}$

The square root of the Mean Square of the Residuals represents the Standard Error of Estimates (S_{yx}), i.e. the average error of the estimates made with the regression equation in relation to the real observations of the dependent variable. It can be represented as a percentage of the mean of the observations, and is then called the Standard Error of Estimates in Percentage (Syx%), also called the Coefficient of Variation (CV) of the regression.

4.1.2 Coefficient of Determination (R²)

> ➢ Linear equations - The coefficient of determination is given by the quotient between the sum of squares of the regression and the sum of squares of the total:

$$R^2 = SQ_{reg} / SQ_{total}$$

Where: R^2 = coefficient of determination; SQ_{reg} = regression sum of squares; SQ_{total} = total sum of squares.

> ➢ Non-linear equations - The coefficient of determination is given by the difference of unity and the ratio between the sum of squares of the residuals and the sum of squares of the total:

$$R^2 = 1 - (SQ_{res} / SQ)_{total}$$

Where: R^2 = coefficient of determination; SQ_{res} = sum of squares of the residue; SQ_{total} = sum of total squares.

4.1.3 Adjusted Coefficient of Determination (R²)aj

When necessary, in the case of a different number of parameters between the best models, or when one model is linear and the other non-linear, or when two linear models are compared and one of them has the constant b_0 and the other does not, the R^2 should be adjusted to enable the models to be compared and the best one to be chosen from the equation:

$$R^2{}_{aj.} = R^2 - \left(\frac{k-1}{N-k}\right) . (1 - R^2) = R^2 - \left(\frac{GLreg}{GLres}\right) . (1 - R^2)$$

Where: $R^2{}_{aj.}$ = adjusted coefficient of determination; k=*number* of equation coefficients; N=*number* of observations; GLreg=degrees of freedom of the regression; GLres=degrees of freedom of the residual.

4.2 Validation of regression equations

The tests used to validate the regression equations can be carried out using the SAS System statistical package according to the procedures described by SAS (2004).

The selected equations are validated by determining:

- ➢ Homoscedasticity of variance using the χ^2 test;
- ➢ The independence of residuals using the Durbin-Watson test or the Breusch-Godfrey test;
- ➢ The normality of the distribution of the residuals using the Kolmogorov-Smirnov or Shapiro-Wilk tests.

4.2.1 Homoscedasticity of variance

One of the main assumptions for the usual least squares regression is homogeneity of variance (homoscedasticity). If the model is well-fitted, there should be no pattern to the residuals plotted against the fitted values. If the variance of the residuals is not constant, then the residual variance is said to be "heteroscedastic". There are graphical and non-graphical methods for detecting heteroscedasticity. A commonly used graphical method is to plot the residuals against the fitted values, as described in section "3.9 Criteria for selecting regression equations". The SAS System calculates the residuals and fitted values using the GLM, REG and NLIN procedures, which can be presented in a graph. When the residuals are distributed without any pattern, there is no heteroscedasticity.

A mathematical method for determining whether there is homogeneity of variance in the residuals, which can be carried out using the SAS system, is White's test (SAS, 2004). White's test is computed by finding nR^2 from a regression of e_i^2 on all distinct variables in $X \times X$, where X is the vector of dependent variables including a constant. This statistic is asymptotically distributed as Chi-squared (χ^2) with k-1 degrees of freedom, where k is the number of regressors.

The method tests the null hypothesis that the residual variance is homogeneous. Then, if the "p" value is too small, the hypothesis is rejected and the alternative hypothesis that the variance is not homogeneous is accepted. This is done by using the "SPEC" option in the model statement, as shown in the following example:

```
PROC REG;
  MODEL Y = X / SPEC;
```

The test can also be run using the WHITE option in the FIT statement of the MODEL procedure in SAS, as in the following example:

```
PROC MODEL;
  PARMS A B C;
  Y = A + B * X1 + C * X2;
  FIT Y / WHITE;
```

4.2.2 Independence of residues

4.2.2.1 Durbin-Watson test

The Durbin-Watson test evaluates errors of the AR(1) type. The value of the Durbin-Watson "d" statistic (SAS, 2004) is obtained using the CLM option in the MODEL statement of the SAS GLM procedure, or the DWPROB option in the FIT statement of the MODEL procedure, or the DW option in the MODEL statement of the REG procedure, as shown in the example below:

```
PROC REG;
  MODEL Y=X1 X2 / DW DWPROB;
```

The "d" statistic is expected to be approximately equal to 2 if the residuals are independent. Otherwise, if the residuals are positively correlated, it will tend to be close to 0 (zero), or close to 4 if the residuals are negatively correlated (Nemec, 1996).

The value of d is given by:

$$d = \frac{\sum_{i=2}^{n}(E_i - E_{i-1})^2}{\sum_{i=1}^{n} E_i^2}$$

Where: d = Durbin-Watson "d" statistic; E_i = stochastic error = $\hat{Y}_i - Y_i$; n = number of observations; $\hat{Y}_i$ = estimated value; Yi = observed value.

4.2.2.2 Breusch-Godfrey test

The Breusch-Godfrey test considers the possibility of ARMA(p,q) type errors, with errors of order 1 to n. The test is available in SAS as an option in the FIT statement of the MODEL procedure, as in the example:

```
PROC MODEL DATA=DATA;
PARMS A=1 B=1;
Y=A*X**B;
FIT Y / GODFREY=3; *TESTS THE AUTOCORRELATION OF
ERRORS OF ORDER 3;
```

4.2.3 Normality of the distribution of residuals

The principle of this test is based on comparing the cumulative frequency curve of the data with the theoretical distribution function hypothesized. When the two curves overlap, the test statistic is calculated as the maximum difference between them. The magnitude of the difference is established according to the probability distribution of this statistic, which is tabulated. If the experimental data deviates significantly from what is expected from the hypothesized distribution, then the curves obtained should be equally distant and, by analogous reasoning, if the fit to the hypothesized model is admissible, then the curves develop closely.

The Kolmogorov-Smirnov statistic (D) (SAS, 2004) is an Empirical Distribution Function (EDF) statistic. The Empirical Distribution Function (EDF) is defined for a set of n independent observations $X_1, \ldots, X_n$ with a common distribution function F(x). Under the null hypothesis, F(x) is the normal distribution. The observations are ordered from smallest to largest as X(1),..., X(n).

The empirical distribution function Fn(x) is defined as:

$$F_n(x) = 0, \; x < X_{(1)}$$
$$F_n(x) = i/n, \; X_{(i)} \leq x < X_{(i+1)}, \; i = 1,2,\ldots,n\text{-}1$$
$$F_n(x) = 1, \; x_{(n)} \leq x$$

Fn(x) is a sequential function that advances by [1/n] with each observation. This function calculates the distribution function F(x). At any value x, Fn(x) is the proportion of observations less than or equal to x, while F(x) is the probability of an observation being less than or equal to x. EDF statistics measure the discrepancy between Fn(x) and F(x). The computational formulas for EDF statistics make use of the transformation of the probability integral U=F(X). If F(X) is the distribution function of X, the random variable U is uniformly distributed between 0 and 1.

Given n observations of X(1),..., X(n), the values U(i)=F(X(i)) are computed as shown below. The Kolmogorov-Smirnov statistic (D) is based on the largest vertical difference between F(x) and Fn(x), and is defined as:

$$D = \sup_x |F_n(x) - F(x)|$$

The Kolmogorov-Smirnov statistic is computed as the maximum of D+ and D-, where D+ is the greatest vertical distance between the EDF and the distribution function when the EDF is greater than the distribution function and D- is the greatest vertical distance when the EDF is less than the distribution function.

$$D^+ = max_i \left((i/n) - U \right)_{(i)}$$
$$D^- = max_i \left(U_{(i)} - (i-1)/n \right)$$
$$D = max \left(D^+, D^- \right)$$

The SAS CAPABILITY procedure uses the modified Kolmogorov D statistic to test the data against the normal distribution with mean and variance equal to the sample mean and variance. In the MODEL procedure the statistic is only used for samples over 2000 individuals. In the case of small samples, the Shapiro-Wilk test described in section "3.8 Data normality test" is used instead of the Kolmogorov-Smirnov test. In the MODEL procedure, the normality test is obtained using the NORMAL option in the FIT statement, as in the example:

```
PROC MODEL;
  PARMS A B C;
  Y=A+B*X1+C*X2;
  FIT Y / NORMAL;
```

5 MORPHOMETRY

The study of the morphometric relationships of the crown and trunk of trees gives an indication of the situation of trees in forest stands and is important for decision-making in forest management actions.

5.1 Cup shapes

Trees have two main types of crown in relation to the type of growth, monopodial or sympodial, and each can have a natural or forest form (Figura 27).

In sympodial growth, the apical bud dominates over the others, forming a single trunk with lateral branches that are not very prominent.

In sympodial growth there is no dominance of one bud over the other and the branches are vigorous from a certain point, where the trunk splits and forms the crown.

Trees that grow in the midst of others in a wood or forest generally have a so-called forest shape, with greater height, thinner diameter and narrower crowns than those that grow free from competition in open environments. Trees growing in open environments have a natural shape, greater crown length and radius, thicker trunk diameters and are shorter than those growing in a forest environment.

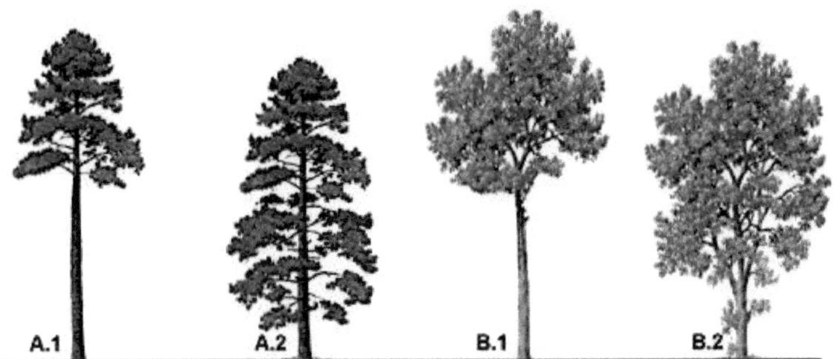

Figura 27 - A) Monopodial growth, forest form (A.1), natural form (A.2); B) Sympodial growth, forest form (B.1), natural form (B.2). Source: Imaña et al (2002).

5.2 Morphometric relationships

5.2.1 Degree of slenderness

The degree of slenderness is an indicator of the trunk's cylindricity; the higher it is, the greater the cylindricity, which can indicate excess population density over long periods. It is calculated using the equation:

$$GE = h / d$$

Where: GE = degree of slenderness; h = tree height in meters; d = trunk diameter in centimeters at 1.3 m height.

5.2.2 Cup formal

The crown shape (FC = DC / CC) is the corresponding crown shape to the degree of slenderness of the trunk. More elongated canopies are related to greater competition between trees, but if it is a natural characteristic of the species, it can also be an indicator that the species is more efficient at occupying space, as a more elongated canopy receives more light than a more rounded one.

5.2.3 Comprehensiveness index

-Breadth   Index (AI = DC / h) is an indicator of space occupation as a function of height; broader canopies mean that trees have little competition with each other.

5.2.4 Overhang index

-The   Overhang Index: (IS = DC / d), also known as the Seebach Growth Space Factor, or crown projection ratio, shows how much the crowns are contributing to the thickening of the tree trunk; the lower the IS, the more the trunk is thickening with a smaller crown diameter.

5.2.5 Living space index

-The   Living Space Index (LIVI = DC^2 / d^2) is also known as the ground cover area quotient and shows how much living space is occupied by the tree in relation to its basal area.

-The   study of minimum and maximum VSI limits by species and site class makes it possible to determine the minimum and maximum surface area to be occupied by a tree (ASSMANN, 1970).

-Average number of   trees per hectare (N) as a function of the space occupied by the tops of trees growing free of competition (ASSMANN, 1970):

-In   square spacing: $N = 10000 / DC^2$;

-In   triangular (or hexagonal) spacing:

$$10000 / (0.866.DC^2);$$

-In   mixed spacing (between square and hexagonal):

$$10000 / (0.933.DC^2).$$

-The   crown projection surface (CPS) of an individual tree can be referred to as approximately proportional to the growth space or nominal area occupied by this tree (SEEBACH apud Assmann, 1970).

5.3 Crown morphology

The shape of the canopy varies from species to species and within the same species between the different genetic strains. The size of the crown is extremely important for growth.

The canopy usually has an upper part exposed to sunlight with a geometric shape between a cone and a paraboloid. And another part that is shaded below, which can have different shapes depending on the species. The curve of the part of the crown exposed to the sun can increase with age. Photosynthetic efficiency differs between the two parts. In the morphological study of the crown, it is necessary to model the part exposed to the sun and the shaded part separately.

The greater the surface area of the canopy that is exposed to the sun, the more efficient the tree is at photosynthesizing and growing. However, the volume and surface area of the tree canopy are calculated as approximations, as precise calculations are impossible (ASSMANN, 1970 (p.115)).

BURGER (1939) apud Assmann (1970) studied the morphology of the *Picea* canopy, creating a pattern that can be applied to most tree species, as reproduced in Figura 28.

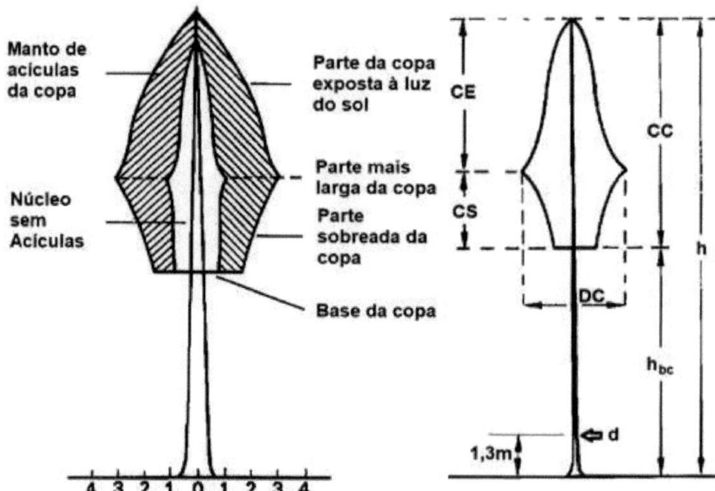

Figura 28 - Picea canopy morphology. Where: CE = crown length in sunlight; CS = crown length in shade; DC = crown base width = crown diameter; CC = crown length; hbc = crown base height; h = tree height; d = tree diameter. Sources: BURGER (1939) apud Assmann (1970).

The morphology of the canopy changes depending on the competition received from nearby trees. Trees that receive little competition tend to have wider and more voluminous canopies and this depends very much on the position of the canopy in relation to the canopy.

5.4 Classification of trees according to their position in the canopy

The classification of trees with reference to the position of the crown in the canopy usually includes the classes shown in Figura 29 and listed below:

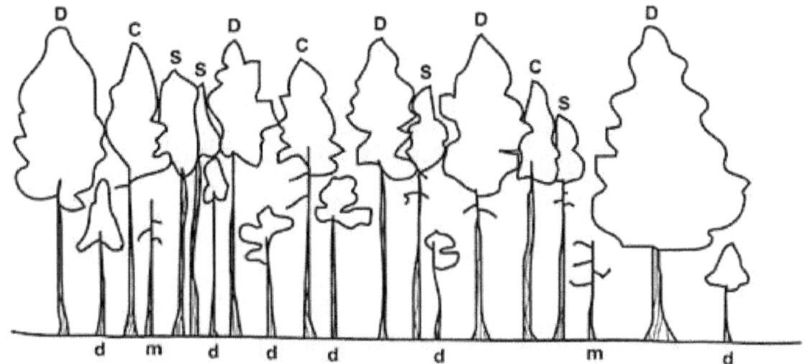

Figura 29 - Class of trees in relation to the position of their crowns in the canopy: dominant (d); codominant (c); subdominant or intermediate (s); dominated or suppressed (d); dead (m). Source: adapted from Alves, Pereira and Correia (2012).

➢ Emergent - These are trees whose trunks and crowns are well formed, with the number and size of branches proportional to their size; their crowns are above the canopy, even above the dominant ones; they receive full light from above and from the sides; their occurrence is very small and they are generally selected for entry into breeding programs.

➢ Lobas - These are trees that develop and grow free of competition; they have a thicker trunk than normal; the branches are thick and numerous on all sides, with branches well below the level of the canopy; their canopies receive direct light from above and largely from the sides; sometimes they are forked.

➢ Dominant - Trees with a crown extending above the canopy that receive full sun from above and partly from the sides. The crowns are well developed and wider than the canopy.

➢ Codominant - These are trees with canopies that form the level of the canopy and receive full light from above, but little from the sides; their canopies have a medium radius;

➢ Subdominant or Intermediate - These trees are lower than the previous classes, have a smaller crown diameter and their tops receive direct light, but none from the sides.

➢ *Overtopped* - These are trees whose crowns are below the level of the canopy and do not receive direct light. Overtopped trees tend to die from excessive competition and some of the other classes may die from disease or pest attacks, or even from being struck by natural phenomena such as lightning.

➢ Dead - These are dead trees, regardless of their position in the canopy.

6 TORSO SHAPE, PROFILE AND VOLUME

The main components of trees are the root, the trunk and the crown (Marchiori, 2004). The stem is the commercial part of the tree trunk and is the main object of study for quantifying commercial wood (Ormond et al., 2006). The study of shapes, volumes and bark is important for quantifying forest production. Timber production planning depends on this information, which is considered critical to the success of timber forestry ventures.

6.1 Trunk shape

The shape of the trunk varies according to the type of growth of the tree species, which can be monopodial or sympodial. The trunk of sympodial trees ends abruptly where the branches of the crown begin. Monopodial trees have a single trunk from base to apex.

Most trees can have their trunk represented by a geometric figure. There are basically 4 types of geometric figures to represent the trunk: cylindrical, paraboloid, conical and neiloid (Figura 30).

The shape of the trees can be estimated using the equation:

$$g_x = p.x^r$$

Where: g_x = basal area (m²); x = distance (m) from the top; p, r = coefficients (p = size; r = shape).

When the coefficient r of the equation is close to 1, the shape of the trunk is cylindrical, close to 0.5 it is paraboloid, close to 1/3 it is conical and below 1/3 it is neiloid (Figura 30).

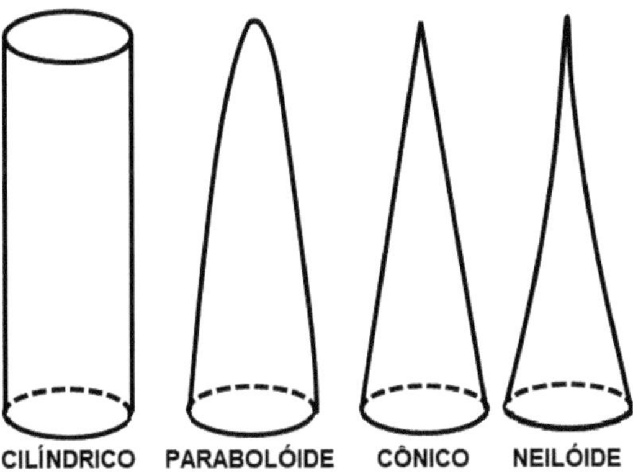

Figura 30 - Geometric shapes resembling tree trunks.

The volume is calculated by integrating the equation as:

$$v = \int_{x=0}^{h} g_x \cdot d_x$$

Where: v = volume of the solid of revolution; gx = basal area (m²); x = distance (m) from the top; h = tree height; p and r = coefficients (p - size; r - shape).

The trunk of trees can have a general shape, but sections of the trunk generally have different shapes. The trunk profile of trees with monopodial growth generally follows a sigmoid curve, being neiloid at the base, cylindrical above the DBH, paraboloid higher up and neiloid at the top. Sympodial trees tend to be neiloid at the base, cylindrical at mid-height, paraboloid just above and with a truncated top where the crown branches begin (Figura 31).

93

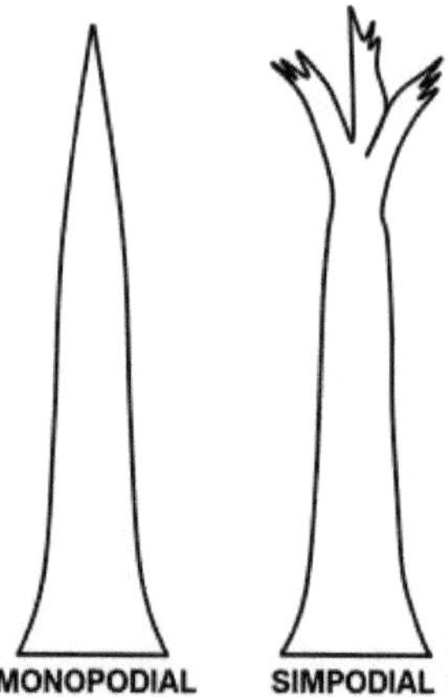

Figura 31 - Tree trunks with monopodial and sympodial growth.

6.2 Strict volume

The Huber, Newton, Smalian, Hossfeld, Simony and Hohenadl equations make it possible to obtain the strict volume of trees and are expressed as follows:

$$Huber\ v = v_0 + \sum L_i\ g_{mi} + v_n$$

$$Smalian\ v = v_0 + \sum L_i\ (\ g_i + g_{i+1}\ ) / 2 + v_n$$

$$Newton\ v = v_0 + \sum L_i\ (\ g_i + 4\ g_m + g_{i+1}\ ) / 6 + v_n$$

$$Hossfeld\ v = v_0 + \sum L_i\ (\ 3\ g_{1/3} + g_{i+1}\ ) / 4 + v_n$$

$$Simony\ v = v_0 + \sum L_i\ (\ 2\ g_{1/4} - g_{mi} + 2\ g_{3/4}\ ) / 4 + v_n$$

Where: v = volume of the tree; v_0 = volume of the stump; v_n = volume of the tip; L_i = length of section i of the tree trunk; g_{mi} = cross-sectional area in the middle of section i; g_i = cross-sectional area at the base of section i; g_{i+1} = cross-sectional area at the top of section i; $g_{1/3}$, $g_{1/4}$ and $g_{3/4}$ = cross-sectional area at 1/3, 1/4 and 3/4 of the base of section i of the trunk; i = order number of the trunk section.

The calculation of the volume of the stump (v_0), except for the Hohenadl equation, is carried out using the equation:

$$v_0 = L_0 \, g_0$$

Where: v_0 = volume of the stump; L_0 = length or height of the stump; g_0 = cross-sectional area of the stump at the top.

In sympodial trees, the volume of the tip, which is the last log, is calculated in the same way as the other logs, while in monopodial trees, the volume of the upper tip of the trunk is calculated as a cone, according to the equation:

$$v_n = L_n \, g_n \, / \, 3$$

Where: v_n = trunk tip volume in trees with a monopodial trunk; L_n = tip length; g_n = cross-sectional area at the base of the tip.

The Hohenadl method involves measuring the trunk diameters at 10% of the height ($d_{0,1h}$), 30% of the height ($d_{0,3h}$), 50% of the height ($d_{0,5h}$), 70% of the height ($d_{0,7h}$) and 90% of the height ($d_{0,9h}$); the average cross-sectional area corresponding to the 5 diameters is then calculated and multiplied by the height of the tree to obtain the volume, as follows:

$$v = h \, (g_{0,1h} + g_{0,2h} + g_{0,3h} + g_{0,4h} + g_{0,5h}) \, / \, 5$$

, or

$$v = h \, (\pi \, d_{0,1h}{}^2 / 4 + \pi \, d_{0,2h}{}^2 / 4 + \pi \, d_{0,3h}{}^2 / 4 + \pi \, d_{0,4h}{}^2 / 4 + \pi \, d_{0,5h}{}^2 / 4) / 5$$

, or

$$v = h \, \pi \, (d_{0,1h}{}^2 + d_{0,2h}{}^2 + d_{0,3h}{}^2 + d_{0,4h}{}^2 + d_{0,5h}{}^2) \, / \, 20$$

Where: v = trunk volume; h = tree height; $g_{0,1}$, $g_{0,3}$, $g_{0,5}$, $g_{0,7}$, $g_{0,9}$ = trunk cross-sectional area at 10%, 30%, 50%, 70% and 90% of tree height (h) from the base to the top.

The measurement of the trunk to determine the rigorous volume by the Huber, Newton and Smalian methods is carried out as in Figura 32.

The Smalian volume equation is officially adopted in Brazil to determine the volume of logs.

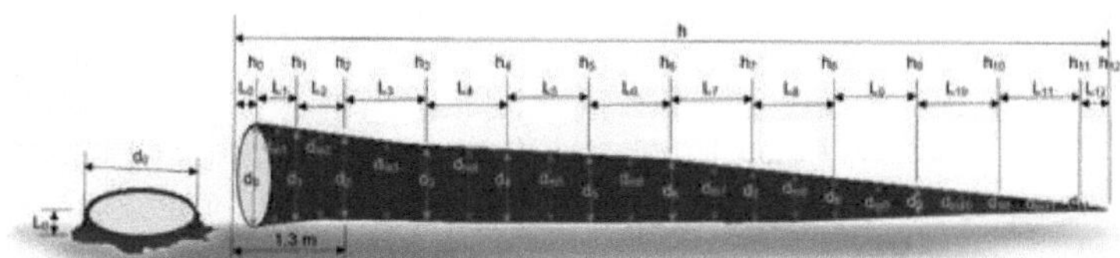

Figura 32 - Measurements to determine the trunk volume of a tree .

Considering the Figura 32 with the measurements and calculations in Tabela 2the volumes calculated by the Huber, Smalian and Newton methods show small differences.

TABELA 2 - Measurements and calculations of tree volume using the Huber, Smalian and Newton methods.

Section no.	h_i m	L_i m	d_i (cm)	d_{mi} (cm)	g_i (m²)	g_{mi} (m²)	$(g_{i-1} + g_i)/2$ (m²)	$(g_{i-1} + 4 g_{mi} + g_i)/6$ (m²)	Huber (m³)	Smalian (m³)	Newton (m³)
0	0,35	0,35	19,6		0,03017		0,03017	0,03017	0,01056	0,01056	0,01056
1	0,95	0,60	15,9	16,5	0,01986	0,02138	0,02501	0,02259	0,01283	0,01501	0,01356
2	1,30	0,35	14,3	14,8	0,01606	0,01720	0,01796	0,01745	0,00602	0,00629	0,00611
3	2,50	1,20	12,8	13,4	0,01287	0,01410	0,01446	0,01422	0,01692	0,01736	0,01707
4	3,70	1,20	12,6	12,9	0,01247	0,01307	0,01267	0,01294	0,01568	0,01520	0,01552
5	4,90	1,20	11,7	11,9	0,01075	0,01112	0,01161	0,01128	0,01335	0,01393	0,01354
6	6,10	1,20	10,5	11,2	0,00866	0,00985	0,00971	0,00980	0,01182	0,01165	0,01176
7	7,30	1,20	9,2	9,6	0,00665	0,00724	0,00765	0,00738	0,00869	0,00918	0,00885
8	8,50	1,20	7,2	8,2	0,00407	0,00528	0,00536	0,00531	0,00634	0,00643	0,00637
9	9,70	1,20	5	5,9	0,00196	0,00273	0,00302	0,00283	0,00328	0,00362	0,00339
10	10,90	1,20	2,9	3,8	0,00066	0,00113	0,00131	0,00119	0,00136	0,00157	0,00143
11	12,10	1,20	1,1	1,8	0,00010	0,00025	0,00038	0,00030	0,00031	0,00045	0,00035
12	12,53	0,43							0,00001	0,00001	0,00001
Total (v)									0,10717	0,11127	0,10854

Where: v_i = volume of the trunk section of order i; L_i = length of section i of the tree trunk; g_{mi} = cross-sectional area in the middle of section i; g_i = cross-sectional area at the base of section i; g_{i+1} = cross-sectional area at the top of section i; $g_{1/3}$, $g_{1/4}$ and $g_{3/4}$ = cross-sectional area at 1/3, 1/4 and 3/4 of the base of section i of the trunk; i = order number of the trunk section.

Newton's equation is considered the most accurate of the three equations used here. The volume of this tree calculated using Huber's equation is 1.26% lower than Newton's and the volume calculated using Smalian's equation is 2.52% higher. The Huber equation overestimates the volumes of trees with a parabolic trunk profile and underestimates those with a neiloid profile, which is what happened with the tree in this example (Tabela 2). Unlike the Huber equation, the Smalian equation overestimates the volume of trees with a neiloid profile, as in this example, and underestimates the volume of trees with a parabolic profile. The greatest differences occurred in the first 2.5 meters of height, concluding that the shorter the sections in this part of the trunk, the more accurate the calculated volumes will be.

The volume can be expressed as v_d, where d is the diameter limit for a given assortment of wood; for example: $v_{7,0}$ is the volume of the trunk up to a minimum diameter of 7.0 cm, which may be the limit for pulp production. The index in this case should be expressed to one decimal place so as not to confuse it with age.

When the volume is expressed for a given age, the symbol v_t is used, where t is the age in years, shown in whole numbers; example v_{15} is the volume at 15 years of age.

When the individual volume includes the trunk and branches of the tree, it is represented with the acronym v_b, if it is only the branches, the symbol is v_a. And in the case of the volume of the trunk plus the branches above 7 cm in diameter, the symbol should be v_{7b}. If it is only the volume of branches above 7 cm in diameter, the symbol is $v_{.7a}$

6.3 Log taper

Taper is the reduction in trunk diameter from the base to the top of the tree in millimeters per meter of length (Figura 33).

The taper of the logs is calculated by:

$$CN = (d_n - d_{n-1}) / L$$

Where: CN = log taper (mm/m); d_n = diameter at the thickest point; d_{n-1} = diameter at the thinnest point; L = log length.

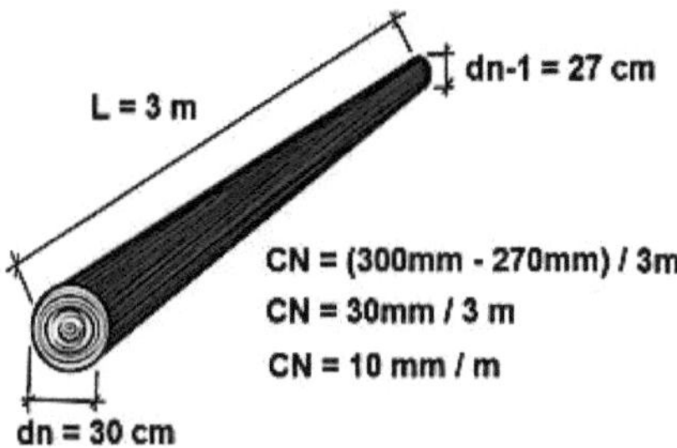

Figura 33 - Determining the taper of a log.

The taper generally has the following characteristics:

➢ It is higher in the logs at the base of the trunk, depending on the species;

➢ It decreases with age;

➢ Within the same species, genetic variation has little influence on conicity;

➢ It is higher in trees that grow more free from competition;

➢ Reduces lumber yield when it is greater than 10 mm/m.

6.4 Trunk profile (tapering and assortments)

Many mathematical models have been proposed to represent the log profile with the aim of separating the different assortments of wood logs by minimum and maximum diameters according to industrial utilization and different log prices on the market. One model that has been used in most cases is Prodan's 5th degree polynomial (MÜLLER, 2004), expressed as follows:

$$d_i = d\,[b_0 + b_1(h_i/\square)/h + b_2(h_i/h)^2 + b_3(h_i/h)^3 + b_4(h_i/h)^4 + b_5(h_i/h)^5]$$

Where: d = diameter at a height of 1.3 m; h = tree height; di = diameter at height hi; hi = height at position i; i = order number of the measurement position; b_0 , b_1 , b_2 , b_3 , b_4 , b_5 = coefficients to be adjusted.

E, whose integral for calculating the volume up to a given height hi is given by (MÜLLER, 2004):

$$V = K.\left[\begin{matrix} a_0^2.h_i + \dfrac{a_1^2.h_i^3}{3} + \dfrac{a_2^2.h_i^5}{5} + \dfrac{a_3^2.h_i^7}{7} + \dfrac{a_4^2.h_i^9}{9} + \dfrac{a_5^2.h_i^{11}}{11} + \\[2ex] + a_0.a_1.h_i^2 + \dfrac{2.a_0.a_2.h_i^3}{3} + \dfrac{a_0.a_3.h_i^4}{2} + \dfrac{2.a_0.a_4.h_i^5}{5} + \dfrac{a_0.a_5.h_i^6}{3} + \\[2ex] + \dfrac{a_1.a_2.h_i^4}{2} + \dfrac{2.a_1.a_3.h_i^5}{5} + \dfrac{a_1.a_4.h_i^6}{3} + \dfrac{2.a_1.a_5.h_i^7}{7} + \dfrac{a_2.a_3.h_i^6}{3} + \\[2ex] + \dfrac{2.a_2.a_4.h_i^7}{7} + \dfrac{a_2.a_5.h_i^8}{4} + \dfrac{a_3.a_4.h_i^8}{4} + \dfrac{2.a_3.a_5.h_i^9}{9} + \dfrac{a_4.a_5.h_i^{10}}{5} \end{matrix}\right]_0^{h_i}$$

Where: v = volume from height 0 to h_i ; K = 10000 (π / 4); a =b_{00} .d; a =b_{11} .(d/h); a =b_{22} .(d/h^2); a =b_{33} .(d/h^3); a =b_{44} .(d/h^4); a =b_{55} .(d/h) .^{5}

The use of the integral is not mandatory, although it is recommended. You can estimate the diameters at certain heights and separate the sections, calculating their volumes using the Smalian equation, for example, or the volume equation used by a particular company.

Müller (2004) worked with nineteen trunk tapering models to describe the trunk profile of *Eucalyptus grandis*, and one of the best models was Prodan's 5th degree polynomial.

The model for finding the height (h_i) at which a given diameter occurs (d_i), in order to separate the trunk sections into assortments, is as follows:

$$h_i = h\,[\,c_0 + c_1(d_i/d) + c_2(d_i/d)^2 + c_3(d_i/d)^3 + c_4(d_i/d)^4 + c_5(d_i/d)^5\,]$$

Where: h_i = height at position i; h = tree height; d_i = diameter at height h_i ; d = tree diameter at 1.3 m height; i = measurement order number; c_0 , c_1 ,c_2 , c_3 , c_4 , c_5 = coefficients to be adjusted.

Consider the set of cubing data for five 14-year-old *Eucalyptus urograndis* hybrids listed in Appendix A, in order to adjust Prodan's model.

Prodan's model for finding the diameter of the trunk d_i at a height h_i is as follows:

$$d_i\,/d = b_0 + b_1\,(h_i\,/h) + b_1\,(h_i\,/h)^2 + b_1\,(h_i\,/h)^3 + b_1\,(h_i\,/h)^4 + b_1\,(h_i\,/h)^5$$

You must first calculate Y, X_1 , X_2 , X_3 , X_4 and X_5 as:

$$Y = d_i\,/d,$$
$$X_1 = (hi/h),$$
$$X_2 = (hi/h)^2,$$
$$X_3 = (hi/h)^3,$$
$$X_4 = (hi/h)\,,^4$$
$$X_5 = (hi/h)\,,^5$$

And then adjust the equation:

$$Y = b_0 + b_1\,X_1 + b_2\,X_2 + b_3\,X_3 + b_4\,X_4 + b_5\,X_5$$

To adjust the model, the variables are converted as follows:

$$h_i/h = c_0 + c_1\,(d_i/d) + c_1\,(d_i/d)^2 + c_1\,(d_i/d)^3 + c_1\,(d_i/d)^4 + c_1\,(d_i/d)^5$$

$$Y = hi/h,$$

$$X_1 = (d_i/d),$$

$$X_2 = (d_i/d)^2,$$

$$X_3 = (d_i/d)^3,$$

$$X_4 = (d_i/d)\,,^4$$

$$X_5 = (d_i/d)\,,^5$$

And then adjust the equation:

$$Y = c_0 + c_1\,X_1 + c_2\,X_2 + c_3\,X_3 + c_4\,X_4 + c_5\,X_5$$

The result is the coefficients shown in Tabela 3.

TABELA 3 - Results of adjusting the equations for trunk taper of 5 *E. urograndis* at 14 years of age

Equation	Coefficients						Statistics			
$d_i/d = f(h_i/h)$	b_0	b_1	b_2	b_3	b_4	b_5	R^2	R^2aj	Syx	Syx%
	1,221057	-4,22035	18,18212	-41,6103	43,27186	-16,9392	0,9817	0,9802	0,0361	5,4%
$h_i/h = f(d_i/d)$	c_0	c_1	c_2	c_3	c_4	c_5	R^2	R^2aj	Syx	Syx%
	1,36495	-3,949915	12,363157	-21,67875	16,21636	-4,25666	0,9874	0,9864	0,0313	6,9%

The first equation determines the diameter at any height position on the trunk and the second determines the height at which a given diameter occurs.

Suppose you have assortments of logs with different prices on the market for energy and for sawmills, with 7 cm being the minimum diameter at a length of 2 m for energy use and 18 cm diameter at 2.7 m for sawmills. What is the height at which the minimum diameter of each assortment occurs for a tree with a diameter of 30 cm and a height of 32 m?

Consider 4 assortments (j), numbered as follows:

- $j = 0$ - stub;
- $j = 1$ - sawmill logs 2.7 m long and at least 18 cm in diameter;
- $j = 2$ - energy logs 2.0 m long and at least 7 cm in diameter;
- $j = 3$ - tip of the trunk considered as waste.

Considering the use for sawmills, with a minimum diameter of 18 cm, the height is calculated as:

$$h_1 = 32 [1.36495 - 3.949915(18/30) + 12.363157(18/30)^2 -$$
$$21.67875(18/30)^3 +$$
$$16,21636(18/30)4 - 4,25666(18/30)5] = 17,08 \ m$$

For energy, with a minimum diameter of 7 cm, the utilization height is calculated as:

$$h_2 = 32 [1.36495 - 3.949915(7/30) + 12.363157(7/30)^2 -$$
$$21.67875(7/30)^3 +$$
$$16.21636(7/30)^4 - 4.25666(7/30)^5] = 28.36 \ m$$

Assuming that the height of the tree as a whole is 0.20 meters and the logs for sawmills (n_j) are sold at 2.7 meters long, the number of logs destined for sawmills is:

$$n_1 = Integer [(17.08 - 0.2) / 2.7) = 6 \ logs$$

Where: n_j = number of logs for assortment j.

And, the utilization height for the sawmill becomes:

$$h_1 = 0.2 + 6 . 2.7 = 16.4 \ m$$

Therefore, the section of the log used for energy goes from 16.4 meters to 28.35 meters in height. The number of logs (n_j) for energy with a length of 2 meters will be:

$$n_2 = Integer\ [\ (28.35 - 16.4)\ /\ 2\ ) = 5\ logs$$

Consequently, the commercial height (h_c) is given by:

$$h_c = h_2 = 0.2 + 6\ .\ 2.7 + 5\ .\ 2 = 26.4\ m$$

The limit heights of the assortments and the number of logs are shown in Tabela 4by assortment.

TABELA 4 - Commercial heights and number of logs of the 5 *Eucalyptus urograndis* at 14 years of age

Assortment	Tree no.	d (cm)	h (m)	h_0 (m)	L_j (m)	d_j (cm)	dj/d	A_j (m)	nj	hj (m)
Energy	1	32,4	37,8	0,15	2,00	7	0,2160	34,15	4	32,15
	2	15,2	27,6	0,10	2,00	7	0,4605	19,09	9	18,10
	3	19,8	32,0	0,15	2,00	7	0,3535	25,14	10	23,15
	4	22,7	33,7	0,15	2,00	7	0,3084	27,69	9	27,15
	5	27,2	34,3	0,10	2,00	7	0,2574	29,64	7	29,10
Sawmill	1	32,4	37,8	0,15	2,70	18	0,5556	22,25	8	24,15
	2	15,2	27,6	0,10	2,70	18	1,1842	0,04	0	0,10
	3	19,8	32	0,15	2,70	18	0,9091	4,38	1	3,15
	4	22,7	33,7	0,15	2,70	18	0,7930	9,25	3	9,15
	5	27,2	34,3	0,10	2,70	18	0,6618	15,52	5	15,10

Where: d = diameter of the tree; h = height of the tree; h_0 = height of the stump; Lj = length of the sections of assortment j; d_j = minimum diameter for assortment j; A_j = Height where the minimum diameter of assortment j occurs; n_j = number of logs in assortment j; h_j = height where the last section of the commercial volume ends, calculated by $h_j = h_0 + n_1 L_1 + n_2 L_{.2}$

Although integration can be used to determine assortment volumes, log transportation legislation and the industry use the Smalian equation to determine log volumes and the result using the integral will be different. Using the tapering equation associated with the Smalian equation will give a result that is closer to what the industry does and to the legal requirement. This also makes computer calculations easier. All you have to do is calculate the diameters at the thin end of the logs and test to see which assortment they correspond to.

The volumes of the assortments (stumps, logs, logs and tops) determined by the Smalian method for the 5 *Eucalyptus grandis* trees in this study are shown in Tabela 5The volumes were calculated using the actual diameters measured in the field and the diameters estimated by the tapering equation.

TABELA 5 - Actual and estimated volumes with the tapering equation of the assortments of 5 *Eucalyptus urograndis* at 14 years of age, calculated using the Smalian method

Tree no.	d (cm)	h (m)	Volume	v_0 (m³)	v_1 (m³)	v_2 (m³)	v_3 (m³)	v (m³)
1	32,4	37,8	Real	0,00043	0,04196	0,00950	0,00197	0,05386
			Dear	0,00046	0,04769	0,00643	0,00215	0,05673
2	15,2	27,6	Real	0,00014	0,00000	0,01835	0,00068	0,01918
			Dear	0,00014	0,00000	0,01850	0,00077	0,01941
3	19,8	32,00	Real	0,00031	0,01565	0,01379	0,00081	0,00165
			Dear	0,00028	0,01184	0,01640	0,00098	0,00165
4	22,7	33,7	Real	0,00032	0,02191	0,01449	0,00046	0,00165
			Dear	0,00032	0,02099	0,01412	0,00058	0,00165
5	27,2	34,3	Real	0,00027	0,02493	0,01441	0,00096	0,00165
			Dear	0,00026	0,03279	0,00949	0,00111	0,00165

Where: d = tree diameter; h = tree height; v_0, v_1, v_2, v_3 = tree assortment volumes (0=timber; 1=sawmill; 2=energy; 3=tip residue); v = total tree volume.

The calculation of volumes using the Smalian method, section by section of the trunk, are shown in Appendices A and B, for the actual and estimated volumes, respectively.

The analysis of variance did not show any significant differences for the volumes of the assortments, nor for the actual volume. As can be seen, the values for the actual and estimated volumes are very similar for all the assortments and for the total volume of the trees. This demonstrates the viability of the procedure adopted in this work, which does not require the use of the integral equation.

6.5 Volume without peel

In cubing, the same deterministic equations of Smalian, Huber, Hohenadl, Newton, etc. are used to find the bark-free volumes from the bark-free diameters. But bark-free volume can also be obtained using stochastic equations, such as the Spurr and Hush model, Stoate, etc., which estimate bark-free volume using independent variables such as tree diameter, height or bark volume itself, or even thinning equations such as the Munro, Kozak, Prodan, etc. models.

Stochastic equation models for estimating bark-free volume can therefore be mainly of the following types:

$$v_s = f(d)$$

$$v_s = f(d, h)$$

$$v_s = f(v)$$

$$v_s = f(d, h, d_{si}, h)_i$$

Where: v_s = volume without bark; d = tree diameter at 1.3 m height; h = tree height; v = volume with tree bark; d_{si} = trunk diameter at position i; h_i = measurement height of trunk diameter of order i.

6.6 Bark volume (vc)

The bark volume (vc) of a tree is usually obtained from the difference between the volume with bark and the volume without bark.

6.7 Commercial volume (v)$_{com}$

The commercial volume (v_{com}) corresponds to the volume of the stem (v_f) of the tree minus the volume of the stump (v_0). The stem volume (v_f) is the volume of the trunk used up to a certain minimum diameter determined by the market, or by the industry that will use the logs.

6.8 Volume of wood for lamination

The volume of wood for veneer production on a lathe is calculated for complete and incomplete veneers and the volume of wood in the cylinder remaining after lamination. Lathes cannot reach all the way to the center of the log, leaving an unlaminated cylinder usually with a diameter of 8 to 11 cm, depending on the lathe.

Figura 34 - A) Log being rolled on a lathe; B) Cylinder remaining after rolling. Source: Vantec (2021).

Wooden logs are generally conical, i.e. one end has a larger diameter than the other, generating incomplete blades until the lathe knife reaches the diameter of the thin end of the log (Figura 35). The volume of wood in the remaining cylinder is calculated according to the lathe's specifications. The volume of wood from complete veneers is calculated by subtracting the volume of the log calculated with the diameter of the thin end, minus the volume of the remaining cylinder and the volume of wood. The volume of incomplete veneers is calculated by subtracting the total volume of the log, minus the volume calculated with the fine tip.

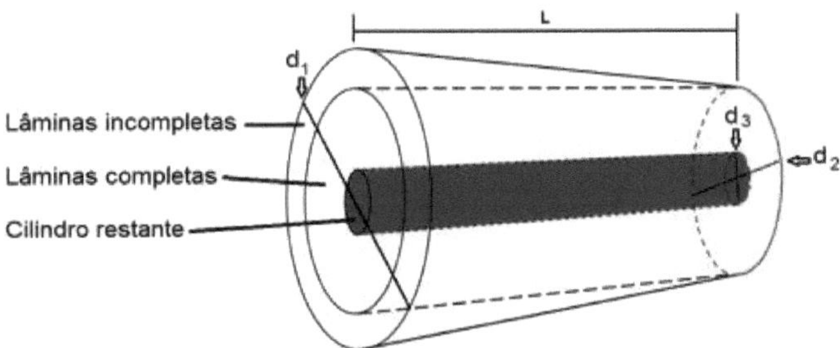

Figura 35 - Calculating the volume of wood for laminating in turning mills.

Being:

> d_1 = diameter of the thick end of the log;

> d_2 = diameter of the thin end of the log;

> d_3 = diameter of the cylinder remaining after rolling;

> v_1 = total log volume calculated using the Smalian method:

$$v_1 = L \cdot (\pi \cdot d_1{}^2 / 4 + \pi \cdot d_2{}^2 / 4) / 2$$

> v_2 = log volume calculated using the diameter of the fine tip:

$$v_2 = L \cdot \pi \cdot d_2{}^2 / 4$$

> v_3 = volume of the remaining cylinder:

$$v_3 = L \cdot \pi \cdot d_3{}^2 / 4$$

> Volume of incomplete slides (v_{LI}) is calculated as:

$$v_{LI} = v_1 - v_2$$

➤ Volume of complete blades (v_{LC}) is calculated as:

$$v_{LC} = v_2 - v_3$$

7 VOLUME OF STACKED WOOD

Stacked wood is measured by the apparent volume of the wood pile in its three dimensions, width, length and height, and the unit of measurement is the stereo, whose symbol is st.

Stereo was officially defined by INMETRO (1999) as: "the volume of a pile of round timber, contained in a cube whose edges measure one meter, including the normal empty spaces between the logs, these spaces being those present in a pile of logs accommodated one to the other longitudinally".

Although the "stereo" measure is still used to measure and sell firewood informally, it is not an official measure in Brazil, having been banned from the official system in favor of the volume in solid cubic meters, used in the international system of measures, as of December 31, 2009 by INMETRO through Ordinance No. 130 of December 7, 1999, reproduced below:

"Ministry of Development, Industry and Foreign Trade-MDIC

National Institute of Metrology, Standardization and Industrial Quality - INMETRO

Ordinance No. 130 of December 7, 1999

THE PRESIDENT OF THE NATIONAL INSTITUTE OF METROLOGY, NORMALIZATION AND INDUSTRIAL QUALITY - INMETRO, in the use of his attributions, conferred by Law no. 5.966, of December 11, 1973, and in view of the provisions of subparagraph "a", of sub-item 4.1, of the Metrological Regulations approved by Resolution no. 11/88, of 12 October 1988, of the National Council for Metrology, Standardization and Industrial Quality - CONMETRO, Whereas the International System of Units - SI does not include the stereo unit of measurement;

Whereas the use of the stereo unit of measurement should be discontinued where it is currently used, and should not be adopted where it is not in use;

Considering that the use of the stereo unit of measurement should be gradually abolished, so as not to cause any damage to the segments involved in the commercialization of roundwood, resolves to issue the following provisions:

Art. 1 The stereo unit of measurement, used in operations involving the production, harvesting, transport and sale of roundwood, whether used as fuel or as an industrial raw material, will be used until December 31, 2009, adopting the following requirements:

I - A stereo is understood to be the volume of a pile of round timber, contained in a cube whose edges measure one meter, including the normal empty spaces between the logs, these spaces being those present in a pile of logs accommodated one to the other lengthwise, as per the annex.

II - The volume of a pile of round timber, measured in stereo, is the volume in cubic meters of the three-dimensional geometric figure in which said timber is contained, in this case represented by a rectangular parallelepiped where the sides are: average length of the logs, height and length of the respective pile, as shown in the annex.

III - When there is a variation in height, this should be measured at various points, adopting the average height for quantifying the pile, as per the annex.

Art. 2 All measuring instruments involved in the commercialization of round wood, used as fuel or industrial raw material, must be submitted to metrological control and meet the minimum specifications established by INMETRO, in order to guarantee their metrological reliability.

Art. 3 As of January 1, 2010, only SI units will be allowed in operations involving the production, harvesting, transport and sale of roundwood used as fuel or industrial raw materials.

Art. 4 This Ordinance shall enter into force on the date of its publication in the Federal Official Gazette.

MARCO ANTONIO A. DE ARAÚJO LIMA

President of INMETRO"

The stereo, therefore, cannot be used as a measure in official documents and the volume of wood piles can have their apparent volume measured, but must be converted into solid volume by a cubing factor (CF), which is the ratio between the solid volume in cubic meters and the apparent volume in stereos. The stacking factor (FE) is the inverse of the cubing factor and is used to convert solid volume into apparent volume.

7.1 Measuring piles of wood logs or logs

Measuring the apparent volume of the wood piles (Figura 36) should be taken by measuring the length at the base and top to determine the average length; the height should be measured at different points and its average calculated; the width of the pile is calculated by the average length of the logs; the apparent volume of the pile in stereos is obtained by multiplying the average length by the average height and the average length of the logs.

Whether it's on the ground or loaded onto trailers, the process for measuring the stereo volume is the same: the average length of the logs (C), the average height of the pile (H) and the average length of the pile (L) are determined. Then the three averages are multiplied and the apparent volume of the wood pile is obtained, i.e. the stereo volume.

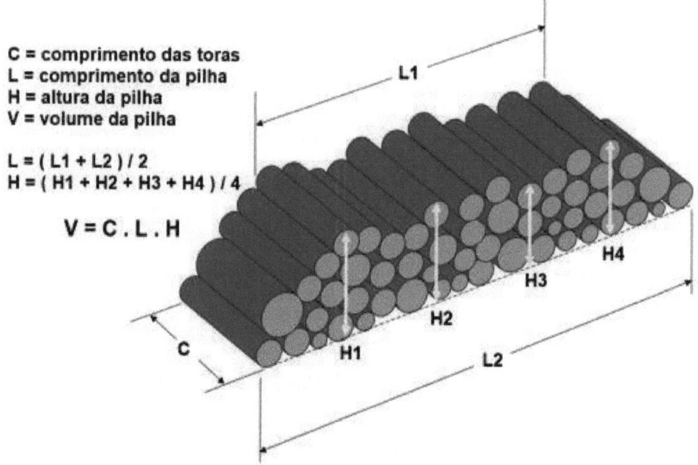

Figura 36 - Measuring the apparent volume of wood piles .

Obtaining the solid volume in m³ from the apparent volume in stereos (st) can be done using different methods, such as:

> Sampling method by reticulated template;
> Xylometer method;
> Angle numbering method;
> Weighing method and density by moisture content;
> Laser beam scanning method.

7.2 Cross-linked template sampling method

In this method, a reticulated template like the one in Figura 37. The template is placed on the side of the pile and the points of intersecting rows and columns that lie on the cross-section of the logs are counted; points on empty space are not counted. The proportion of points on the cross-sections of the logs, or Log

Points (PT), and the total Crossing Points (PC) are then determined to determine the Cubing Factor (CF).

Sampling can be carried out directly on the pile, or with a photograph of the pile, virtually superimposing the reticules on it. A significant number of lattices must be sampled, and a statistically representative number of piles from the same forest must be sampled in order to establish an average that is applicable to all piles in the forest, or to all loads of wood. To do this, the mean and variance of the number of Log Points (PT) are calculated with a certain probability of confidence for the mean (generally 95%) in order to determine the number of sampling units (cross-sections) required.

In the example in Figura 37the total number of intersections is 100 points (10 rows x 10 columns), and the number of intersections on log cross-sections is 47 points, resulting in a ratio of 0.47, which is the Cubing Factor (CF). Calculated by:

$$FC = PT / PC$$

Where: FC = Cubing Factor; PC = number of Log Points; PC = number of Crossing Points.

According to the example given:
- ➢ PT = 47;
- ➢ PC = 100;
- ➢ FC = 47 / 100 = 0.47.

Considering the example given, if the pile has an apparent volume of 100 st, the solid volume of wood in the pile would be 47 m³ (V = 0.47 . 100 st = 47 m³).

Figura 37 - Cubication Factor (CF) sampling with reticulated template.

7.3 Xylometer method

This method involves weighing a truckload of logs and determining the density of the wood using a xylometer. The xylometer is a water tank (Figura 38) to measure the volume of wood displaced by water.

Figura 38 - Measuring wood volume using the xylometer method.

The measurement procedure is carried out as follows:

1) When it arrives at the scales, the truck is loaded, resulting in a weight of P;
2) With the truck still on the scale, a sample of wood is removed with a crane and the remaining weight of the loaded truck without the wood sample is recorded, resulting in a weight p;
3) The weight of the sample taken with the crane (pa) is calculated by reducing the total weight (P) minus the weight of the truck loaded without the sample (p): pa = P - p;
4) The crane plunges the wood into the water tank up to a mark on the crane and the volume of the crane loaded with wood (vgc) is recorded;
5) The volume of wood from the sample (va) removed is calculated by reducing the volume of the grappled crane (vgc) minus the volume of the crane itself (vg) up to the dive mark, which is known in advance: va = vgc - vg;

6) The apparent density of the wood (d) is calculated by dividing the weight of the wood sample (pa) by the volume of the wood sample (va): d = pa / va;
7) The truck is unloaded and returned to the scale to obtain the tare weight (weight of the unloaded truck);
8) The weight of the truck's timber load (PM) is calculated by reducing the weight of the loaded truck (P) minus the tare weight: PM = P - tare weight;
9) The timber volume (V) of the truck's timber load is determined by dividing the weight of the timber load (PM) by the density (d): V = PM / d.

Example:

- Weight of the loaded truck (P):

$$P = 17,000 \ kg$$

- Weight of the loaded truck (p) without the wood sample:

$$p = 16,500 \ kg, \ or \ 16.5 \ t$$

- Weight of wood sample (pa):

$$pa = P - p$$
$$pa = 17,000 \ kg - 16,500 \ kg$$
$$pa = 500 \ kg, \ or \ 0.5 \ t$$

- Volume of the empty crane (vg), previously measured on the xylometer (constant value):

$$vg = 0.1 \ m^3$$

- Volume of the loaded crane (vgc) obtained from the xylometer:

$$vgc = 1.3 \ m^3$$

- Volume of the sample (va) taken by the crane:

$$va = vgc - vg$$
$$vc = 0.9\ m^3 - 0.1\ m^3$$
$$va = 0.8\ m^3$$

- Apparent wood density (d):

$$d = pa / va$$
$$d = 0.5\ t / 0.8\ m^3$$
$$d = 0,625$$

- Volume of the truck's timber load (V):

$$V = PM / d$$
$$V = 16.5\ t / 0.625$$
$$\boldsymbol{V = 26.4\ m^3}$$

If the xylometer's measuring ruler is visible to the scale operator, they can carry out the measurements and calculations themselves.

7.4 Angle numbering method

This method is based on the Bitterlich principle, described in Chapter 14. Bitterlich (1984), *apud* Batista and Couto (2002), suggested using the proportionality constant k = 1/100. The construction of an instrument with this constant is carried out with an angle whose distance between the eyepiece and the objective is 5 times the width of the objective, as in Figura 39. The Cubing Factor (CF) is equal to the number of logs (N) whose diameters appear larger than the angle (α) when viewed from a fixed point installed on the side of the pile of logs, divided by 100 (CF = N / 100).

In the example in Figura 4039 logs with a diameter wider than the angle were counted within the circle whose diameter is 10 times the width of the k value. Logs are considered for counting if they are wider than the angle and have their center inside the circle formed with a diameter of 10 times the value of k (Figura 39). The Cubing Factor is equal to the number of logs counted divided by 100 (FC = 39 / 100 = 0.39).

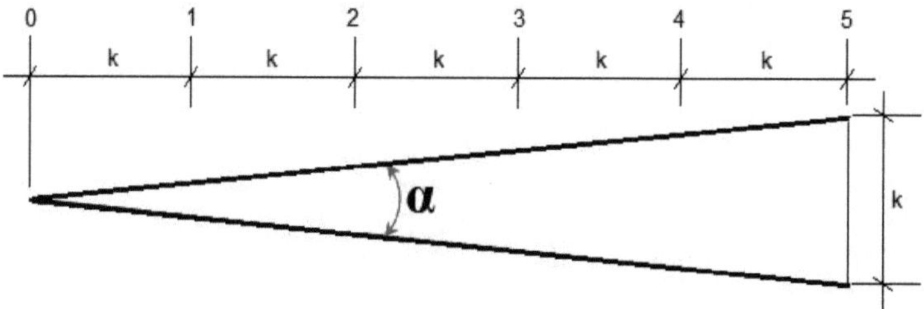

Figura 39 - Angle (α) with an angular constant (k) of 1/100.

Figura 40 - Cubication factor by angle count - the green circles mark the 39 logs counted, with a width greater than the angle of the standard, whose centers are inside the circle with a diameter of 10 times k; CF = 39/100 = 0.39.

7.5 Weighing method and density by moisture content

This method is based on the density of the wood by moisture content. It was widely used in the past, before laser systems were invented. It is necessary to construct a table of basic wood density according to moisture content for each species, or even by clone or genetic strain, in order to achieve greater precision. The loads of wood are weighed and a sample is taken to determine the moisture content and to check the density table by moisture content for the density of the wood and then convert the weight of the load into volume by dividing the weight by the density of the wood.

To obtain the sample and measure the moisture content, it is necessary to cut the log about 30 cm from its top and then use a moisture meter for green wood (Figura 41), which admits values of around 40% to 120% moisture.

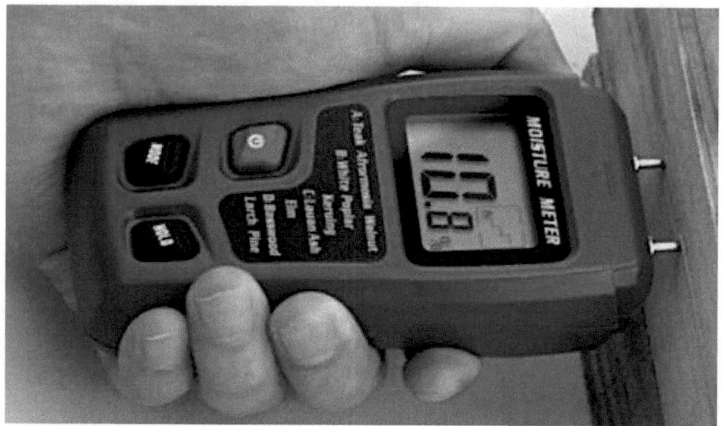

Figura 41 - Moisture meter for green wood. Source: amazon.com (2021).

Using a table as in the example below, check the density corresponding to the moisture content and divide the weight of the load by the density of the wood obtained in the table.

Suppose that the load of logs on a truck weighed 10,000 kg, and the Tabela 6 of wood density as a function of moisture content in percent and that the moisture content sampling resulted in 93% in an average of three logs sampled from the load. The wood density found in Tabela 6 is 0.7857. Therefore, the volume of wood transported will be calculated as 10,000 kg divided by 0.7857, or 12,727.5 m³.

TABELA 6 - Wood density as a function of moisture content (fictitious)

Humidity %	0	1	2	3	4	5	6	7	8	9
40	0,3379	0,3464	0,3548	0,3633	0,3717	0,3802	0,3886	0,3971	0,4055	0,4140
50	0,4224	0,4309	0,4393	0,4477	0,4562	0,4646	0,4731	0,4815	0,4900	0,4984
60	0,5069	0,5153	0,5238	0,5322	0,5407	0,5491	0,5576	0,5660	0,5745	0,5829
70	0,5914	0,5998	0,6083	0,6167	0,6252	0,6336	0,6421	0,6505	0,6589	0,6674
80	0,6758	0,6843	0,6927	0,7012	0,7096	0,7181	0,7265	0,7350	0,7434	0,7519
90	0,7603	0,7688	0,7772	**0,7857**	0,7941	0,8026	0,8110	0,8195	0,8279	0,8364
100	0,8448	0,8533	0,8617	0,8702	0,8786	0,8870	0,8955	0,9039	0,9124	0,9208
110	0,9293	0,9377	0,9462	0,9546	0,9631	0,9715	0,9800	0,9884	0,9969	1,0053
120	1,0138	1,0222	1,0307	1,0391	1,0476	1,0560	1,0645	1,0729	1,0814	1,0898

7.6 Laser beam scanning method.

Scanning logs loaded onto trucks with a laser beam is being popularized by companies such as Woodtech, which developed the Logmeter system.

Logmeter is an automatic system capable of obtaining high-precision volume measurements and estimating the biometric characteristics of timber loaded onto trucks (WOODTECH, 2021). The system uses state-of-the-art laser technology and offers accurate and reliable measurement without direct operator intervention. Logmeter is also supplied with an optional module that allows the measurement of bulk materials loaded into open containers, such as chips, biomass, coal, etc.

Figura 42 -   Scanner that measures wood loads with a laser beam. Source: Woodtech (2021).

8 GROWTH

8.1 Introduction

Wood has always been one of humanity's main inputs, due to the versatility of its uses, which is increasing day by day, and due to the growth and production potential of trees.

The wood is obtained from the tree trunk (Figura 43) and some other woody plants as a result of the biological process of growth, which is influenced by a wide range of genetic and environmental aspects, resulting in different chemical and physical characteristics of the wood for each species and the environment in which they grow, making it possible to use it *in* different situations *in natura* or after processing. We learn how to manipulate genes, manage forests and provide the right environment to obtain wood with characteristics for uses ranging from energy to food, from construction to the production of semiconductors for electronic equipment. Understanding how tree growth works and how we can influence the process is essential to obtaining products with the desired characteristics.

Tree growth is an immensely complex process and not all aspects of wood formation are fully understood (PUNCHES, 2004). Factors that generally have important effects on growth in most plantations are initial spacing and maintenance, silvicultural treatments, thinning and pruning, local conditions (including nutrition) and climatic conditions; in natural forests the main factors are regeneration density and treatments, spatial distribution, silvicultural treatments, artificial suppression of individuals, local conditions and climatic conditions (BRACK and WOOD, 1997).

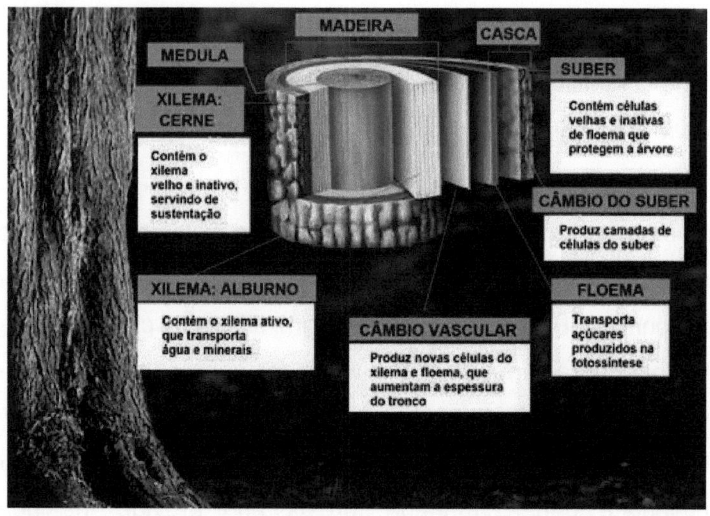

Figura 43 - Anatomical representation of the tree trunk. Source: Bruchez (2017).

The process of tree growth begins with the germination of the seed. The cells produced by the primary apical meristem differentiate to form a tender stem, the first leaves and the secondary meristems that will produce the bark and wood. Under the ground, something similar happens: the root's apical meristem differentiates, initially forming tender roots and the secondary meristems that will produce the cells of the root wood tissues and the bark. The first stem and roots depend on the amount of reserve substances in the seed and serve to begin the process of absorbing water and nutrients and carrying out photosynthesis, so that the secondary meristems can begin their formation and then the production of adult cells.

8.1.1 Phytohormones

Plant hormones control the growth and development of plants, from the beginning of embryogenesis with the meeting of female and male gametes, the regulation of organ size and growth, the response to pathogens, stress tolerance, flowering and the development of reproductive organs, among others. Growth hormones are produced near the apical meristems and are transported through the tree's vascular system (Punches, 2004).

There are five main groups of phytohormones that affect growth: auxins, gibberellins (GA), cytokinins, ethylene and abscisic acid (ABA). The main functions of each are listed below (VANDERZANDEN, 2012):

- ➤ Auxins
  - o The auxins related to thickening, or secondary growth, move to the secondary meristems, through the parenchyma, from the top to the base of the tree; the auxins are produced by the primary meristem itself at the ends of the stem and branches and always move towards the roots, regardless of the action of gravity (Kramer and Kozlowsky, 1972);
  - o They stimulate secondary meristems, promoting the thickening of tree trunks;

- o They regulate phototropism, causing the plant to grow towards the light;
- o Regulates root gravitropism;
- o Promote apical dominance (promotes the production of hormones that suppress the growth of buds below, on the stem);
- o They stimulate the formation of flowers and fruit;
- o They promote the formation of adventitious roots;
- o They stimulate rooting in vegetative propagation (e.g. AIB);

➢ Gibberellins
- o Gibberellins related to elongation, or primary growth, travel little to the dividing cells at the apex of the primary meristem;
- o They stimulate the primary meristem, promoting apical growth (Kramer and Kozlowsky, 1972);
- o They stimulate cell division and elongation;
- o They break seed dormancy and accelerate germination;

➢ Cytokinins
- o Cytokinins are found in plants and animals;
- o They stimulate cell division and are often included in the sterile media used to grow plants from tissue culture;
- o High in cytokinins and low in auxins, it stimulates the production of shoots;
- o A high ratio of auxins to cytokinins stimulates the production of more roots;

o They slow down ageing and death (senescence);

➢ Ethylene gas

o Induces fruit ripening;

o Promotes abscission and senescence;

o In response to stress, ethylene production may be increased;

o The increase in ethylene in leaf tissue is part of the reason why leaves fall off trees;

➢ Abscisic acid

o Abscisic acid (ABA) is a general inhibitor of plant growth;

o Induces dormancy and prevents seeds from germinating;

o It causes leaves, fruit and flowers to wither;

o Regulates the closing of stomata.

TABELA 7 - Relationship between plant hormones and physiological processes

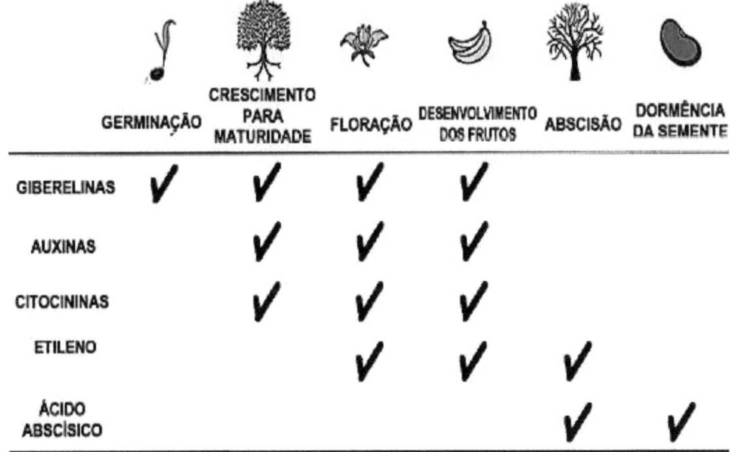

	GERMINAÇÃO	CRESCIMENTO PARA MATURIDADE	FLORAÇÃO	DESENVOLVIMENTO DOS FRUTOS	ABSCISÃO	DORMÊNCIA DA SEMENTE
GIBERELINAS	✓	✓	✓	✓		
AUXINAS		✓	✓	✓		
CITOCININAS		✓	✓	✓		
ETILENO			✓	✓	✓	
ÁCIDO ABSCÍSICO					✓	✓

Source: Biology Online (2021).

8.1.2 Photosynthesis

Growth is the result of cell division and differentiation, increasing the number of cells and consequently the plant's biomass. The wood cells transport raw sap to the green areas (leaves) where photosynthesis (cellulose synthesis), respiration (energy production) and the transformation of nutrients from the raw sap into food for the cells take place. Photosynthetic activity is the main factor in the production of plant biomass. Plants use solar energy to oxidize water ($H_2 O$) and reduce carbon dioxide (CO_2), resulting in one molecule of glucose and six of oxygen, according to the general equation:

$$6\ CO_2 + 12\ H_2\ O \rightarrow C\ H\ O_{6126} + 6\ O_2 + 6\ H\ O_2$$

But to form the cells of higher plants, other processes are needed, which are elaborated in the leaves, using chemical substances absorbed from the soil, air and rainwater diluted in water to form the cytoplasm and all the internal organs of the cells, and glucose from photosynthesis as the basis for forming the cell walls. The process absorbs heat and light. Therefore, plant growth depends on the ambient temperature, insolation, air and water availability, the composition of the air and soil, as well as the texture and structure of the latter, which influence the availability of air and water in it. In addition, the efficiency of each species in growing varies according to the environment in which it evolved and depends on its genetic characteristics, determined by its adaptation to environmental factors over time. Tree growth can be understood as the sum total of the division, elongation and thickening of the cells of the meristematic tissues (Imanã et al., 2005), resulting in an increase in size and biomass, changing the shape of the tree and its parts.

8.1.3 Meristems

The growth tissues of trees, or meristems, are of two types (Herbarium, 2008):

> ➢ Primary - originating directly from the embryo, present at the extremities; they give rise to primary definitive tissues and promote elongation;
> ➢ Secondary - made up of cells that have regained the ability to divide, present between the bark and the xylem of the trunk, branches and roots, of which there are 2 types (Herbarium, 2008):
>> o Felogen - more external, promotes thickening of the bark;
>> o Vascular cambium - more internal, produces xylem cells inwards and phloem cells outwards;

From the activity of the meristems, all the tissues of the wood and other tissues and organs of trees are formed.

The xylem originated by the cambium, as well as transporting raw sap, also supports the crown, consisting of sapwood (physiologically active and less resistant) and heartwood (physiologically inactive, generally lignified, darker and more resistant).

Between the phellogen and the cambium is the secondary phloem, produced by the cambium, which transports elaborated sap and auxins to promote secondary growth.

8.2 Growth conditioning factors

Based on Kramer and Kozlowski (1972), the factors influencing growth were classified as follows:

- Exogenous - referring to or present in the environment;
- Endogenous - referring to the species and individuals cultivated.

8.2.1.1 Exogenous growth factors

These are the environmental variables, or those present in the environment (Kramer and Kozlowski ,1972):

- water availability (topography, soil type, climate);
- availability of light (topography, exposure, latitude);
- heat availability (climate, latitude);
- availability of nutrients (from the soil and fertilizers);
- soil type - the texture and structure of the soil affect:
  - resistance;
  - humidity;
  - aeration;
- competition:
  - tree density of the species;
  - density and aggressiveness of other plant species;
- pests and diseases;

> silvicultural treatments.

8.2.1.2 Endogenous constraints on growth

The main endogenous constraints on growth are (Kramer and Kozlowski,1972):

> - Genetic potential - has to do with vigor and health;
> - Successional stage of the species (tolerance and needs):
>   - pioneers;
>   - initial secondary;
>   - late secondary;
>   - climax;
> - Reserve substances and carbon/nitrogen ratio;
> - Individual dimensions;
> - Age - life stage;
> - Phenological phase and flowering and fruiting season;
> - Type of regeneration:
>   - vegetative (regrowth, cuttings, grafting);
>   - sexed (natural, artificial);
> - Auxin storage in seasonal growth:
>   - ring porosity - some species from temperate regions store auxin from one year to the next;
>   - diffuse porosity - they don't store auxin and start secondary growth only after receiving auxin produced in the apex in early spring - they don't form rings.

8.2.1.3 Seasonal Growth

Species from regions that do not have a well-defined dry or cold season tend to grow continuously, while species from regions with a well-defined dry and rainy or hot and cold season tend to stop growing in the unfavorable season. Species are usually classified according to their growth as follows:

- ➢ Continuous growth (humid tropics);
- ➢ Seasonal growth:
    - o caused by humidity due to the occurrence of a dry season (tropics);
    - o caused by temperature due to the occurrence of a cold season (temperate zones).

Seasonally growing species from temperate zones can have two types of porosity, which can be diffuse or ringed. Those with diffuse porosity start secondary growth after restarting annual auxin production in spring; they do not store auxin from one year to the next and depend on auxin from the apex to start growth. Some ring-porous species from temperate zones store auxin in the wood and can start thickening even before the end of the cold season.

8.3 How trees grow

Trees have two characteristic forms of growth in relation to apical dominance, as follows:

➢ Monopodial growth - there is dominance of the apical bud; the auxins produced in the apical bud inhibit the growth of the lateral branches, forming an elongated crown;

➢ Sympodial growth - there is no dominance of any bud in the crown; there is similar auxin production in the branches and they all grow at similar rates, forming wide crowns.

8.4 Laws of biological growth

According to Imaña-Encinas (2005) the laws of biological growth require a varying amount of critical attention and lose precision and significance as the period of time over which they are applied decreases;

The 5 laws of biological growth of individual organisms are as follows:

1ª - Size is a monotonically increasing function of age;

2ª - The results of biological growth are typically capable of growth themselves.

3ª - In a constant environment, growth occurs at a constant, uniform and specific rate.

4ª - Under current development conditions, the specific acceleration of growth is always negative.

5ª - The specific growth rate declines more and more slowly as the body increases in age.

Growth acceleration is the rate of increase in growth from the beginning to the end of a time interval, expressed as a percentage.

In mathematical terms, growth can be represented by an integral function of the type [$Y = f(t)$]. The first derivative of the function [$Y' = f'(t)$] represents the increase from the beginning to the end of a given period, while the acceleration of growth is represented by the second derivative [$Y'' = f''(t)$] of the growth function.

8.5 Increments

The increment is the increase in the value of the variable studied over a given period of time (Imanã et al., 2005). The increase can be differentiated depending on the time elapsed during which growth is measured (annual, periodic, etc.). The increment value can be determined for an individual or for a population, and can also be expressed per unit area when the variable measured is the same.

All dendrometric variables can have their increment measured; the most common are diameter (d), individual basal area (g), basal area per hectare (G), height (h), dominant stand height (h_{dom}), individual volume (v) and volume per hectare (V).

The main types of increment are: current annual increment; percentage increment, or rate of increment; periodic increment; annual periodic increment; and the average annual increment.

8.5.1 Annual current increase (ACI)

This is the growth between the start of the growing season and the end of the vegetative resting season over a 12-month period, calculated as:

$$ICA = Y_{(m+1)} - Y_{(m)}$$

Where: m = reference year or age; Y = dimension considered.

8.5.2 Percentage increase (p) (Alves, 1982)

$$p = 100 \cdot (Y_{(m+1)} - Y_{(m)}) / Y_{(m+1)}$$

8.5.3 Periodic increase (IP)

This is the total growth over a given period of time:

$$IP = Y_{(m+n)} - Y_{(m)}$$

Where: n = time period; when n = 1 year, IP = ICA.

8.5.4 Annual Periodic Increase (IPA)

This is the average growth per year over a period of time, used for slow-growing species:

$$IPA = (Y_{(m+n)} - Y_{(m)}) / n$$

Where: Y = dimension considered; m = reference year or age; n = time period in years.

8.5.5 Average Annual Increase (IMA)

This is the average growth per year up to the date in question, calculated by:

$$IMA = Y_{(m)} / m$$

Where: $Y_{(m)}$ = Value of the variable at the age considered; m = age.

8.6 Growth and increment curves

The cumulative growth curve usually has a sigmoidal shape and the area under the curve can be represented by an integral function. The first derivative of the integral function expresses the current increase (CI). The second derivative of the integral function expresses the proportion of the current increase from one year to the next.

The growth curve (Figura 44) usually has the following characteristics:

- ➢ An inflection point where the CI is maximum;
- ➢ A maximum tangent where the IM is maximum and is equal to the IC;
- ➢ An asymptote that represents the maximum value it can reach.

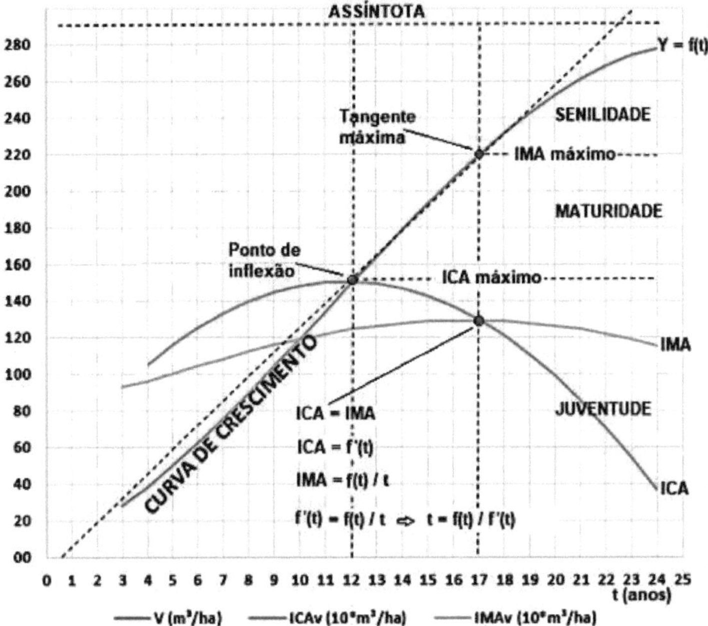

Figura 44 - Growth curve.

8.7 Methods and objectives of studying growth

The growth of a dendrometric variable is a function of genetics, time and the environment in which the tree lives.

If genetics and the environment don't change, growth can be considered a function of time alone:

$$C = f(t)$$

Thus, for the same plant in the same location and considering constant environmental conditions, growth will be an exclusive function of age.

Therefore, the study of growth must take into account the measurement of the variable considered and the time elapsed or age.

Growth is mainly studied for:
- ➢ Forecasting the future and planning production;
- ➢ Assess the quality of the environment;
- ➢ Assessing the genetic quality of individuals and populations;
- ➢ Evaluate the quality of the genotype x environment interaction.

The main methods of measuring variables for studying growth are (IMANÃ et al., 2005):
- ➢ Permanent plots measured annually or periodically;
- ➢ Individual trees identified and measured continuously or periodically (tape);
- ➢ Trunk analysis
  - ○ complete - felled trees;
  - ○ partial - standing trees (Pressler auger).

Age estimates in growth studies can be made by:
- ➢ Observation and recording;
- ➢ Counting verticils of species with monopodial growth - allows calculation of the IMA_d and estimation of all height increments;
- ➢ Growth ring count.

8.8 Applications of growth studies

Some of the main applications of growth studies are:

- ➢ Site classification;
- ➢ Forest stand density management;
- ➢ Weeding planning;
- ➢ Thinning planning;
- ➢ Define the technical rotation age;
- ➢ Production forecast.

8.9 Considerations on growth studies

Tree growth is complex, involving various physiological processes from the assimilation of nutrients to their transformation and growth promotion.

Studying the growth of dendrometric variables is extremely important for characterizing stands and their productive potential.

Comparative growth studies between different genotypes and environments make it possible to select and improve species for each environment and obtain greater productivity with less expenditure of resources of all kinds.

The study of tree growth is therefore essential for forest management and the sustainable management of stands, and serves as a basis for planning all forestry production activities.

8.10 Data collection

Growth data can be obtained from demarcated and duly identified trees in forest stands that are measured periodically, or for those with annual growth rings, samples can be taken in the direction of the trunk radius for later analysis and measurement. In the first case, these can be individual trees, or entire plots in which the trees are permanently numbered and identified. In the second case, the trees can be kept standing and samples taken with an auger, which is called partial trunk analysis, or the trees can be felled and sample wood disks collected along the trunk for later analysis and measurement.

Growth studies can be carried out with the aim of establishing annual growth curves, or growth curves throughout the year to determine when growth is most or least vigorous, or when vegetative growth rests in places where unfavorable seasons occur.

8.10.1 Permanent plots and trees

Permanent plots from continuous inventories or permanent individual trees can be used for growth studies, but the following precautions must be taken:

> ➢ The numbering and location of the plots and trees must be kept unchanged throughout the study, i.e. the same plot will always have the same number over time, and the same tree must also have the same number on all measurement occasions;

- ➢ Measurements must be taken as accurately as possible, using calibrated and reliable instruments;
- ➢ The areas of the sample plots must be subjected to the same silvicultural and management treatments applied to the rest of the stands;
- ➢ Never cut down trees for cubing or other studies within the plots so as not to de-characterize the plots and differentiate them from the rest of the stand;
- ➢ Measurements should preferably be taken in the dormant season and recorded with the full date of the measurement with day/month/year, as it may not be possible to take them in the same period on a given occasion.

8.10.2 Trunk analysis

There are species in regions with unfavorable and favorable seasons for tree growth, resulting in seasonal growth. In spring and summer, they form large cells and lighter tissue, while in fall and winter they form smaller cells and darker tissue, which results in the formation of annual growth rings (Figura 45). This is also due to the existence of a well-defined dry and wet season; in the wet season the cells formed are larger and the tissue is lighter, while in the dry season the cells are smaller and the tissue formed is darker.

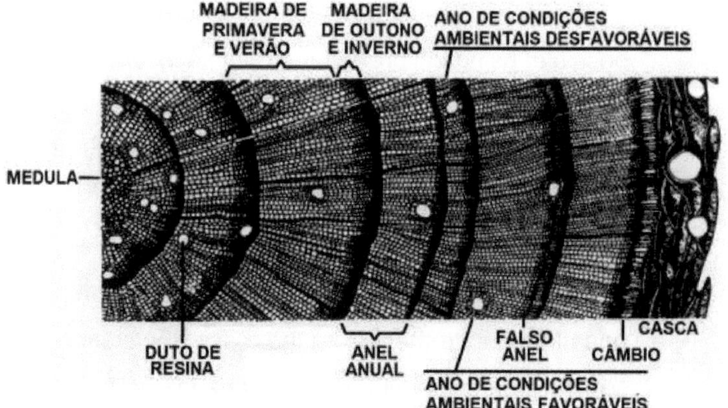

Figura 45 - Typical cross-section of a conifer trunk. Source: Sahin (2012).

Trunk analysis is the technique of studying the growth rings of a tree's trunk. The technique makes it possible to reconstruct the tree's trunk growth history by measuring the diameters each year of the tree's life and inferring the heights reached each year.

Sample trees should be selected using the same procedure as for trees to be cubed, as described in Chapter 13.

Care must be taken not to confuse true annual growth rings with false rings formed for other reasons such as fires, defoliation by insects, disease attacks, unseasonable frosts and droughts, etc. False rings are usually thinner, sometimes incomplete and occur between two true annual rings. It is therefore good practice when carrying out stem analysis to obtain weather data for the region and a history of fires and attacks by pests and diseases in the forest stand.

This technique can be used on standing trees and is called partial trunk analysis (Figura 46), when carried out by extracting cylindrical fillets from the trunk of trees kept alive and standing; and complete trunk analysis (Figura 48), when carried out by extracting sample disks along the trunk of felled trees.

8.10.2.1 Partial trunk analysis

Once the trees have been selected for partial trunk analysis, the following procedure must be followed:

1) Number the tree and write down the species and number;
2) Write down the geographical coordinates of the tree's location;
3) Analyze the trunk at a height of 1.3 meters and identify where the pith should be;
4) Choose the two auger penetration sites at chest level, with one radius at 90° to the other and about 5 cm apart, trying to remove one sample cylinder at about 127.5 cm and the other at about 132.5 cm in height, as there is an indication that if both holes are in the pith, the tree could be killed;
5) Choose the sampling point at the base of the trunk, so that it is minimally higher than half the width of the auger in order to be able to turn it;
6) Measure and note the DBH of the tree with and the thickness of the bark at the points where the rays will be extracted;
7) Measure the height of the tree;
8) Prepare the auger, making sure that the manufacturer's mark is upwards and penetrate the trunk a little further than where you expect to find the pith, but never completely through the trunk;
9) When the penetration stops, try to leave the manufacturer's mark upwards and then insert the extractor to the bottom of the auger, taking care to leave the open part of the extractor upwards;
10) The auger should then be turned back one turn and returned until the manufacturer's mark on the auger is up

again - this will cause the cylinder to break at the end of the penetration, making it possible to remove it;

11)   Start removing the extractor slowly so that you can check that the cylinder is on top of the extractor so that it doesn't fall off when it is removed and that the cylinder has actually broken off at the end of the penetration;

12)   If the cylinder is not loose, repeat steps 9 and 10;

13)   If the cylinder is loose, mark the bark at the top with an ink dot to find out where the sample was on the tree;

14)   Place the cylinder inside a protective cover so that it doesn't get damaged during transportation and identify the sample - two opposing wooden bases can be used, using masking tape to join them together with the wooden cylinder inside;

15)   When you arrive at the laboratory, stick the sample cylinder on a suitable base (Figura 47), in the same position as it was in the tree, i.e. with the mark made during extraction facing upwards;

16)   Allow the sample to air dry or dry in an oven with protection;

17)   Sand the top of the cylinder after the glue has dried;

18)   Mark the position of the rings on the base where the cylinder was glued to facilitate measurement;

19)   Measure and calculate the radii at each age and calculate their measurement proportionally to the ratio of the wet unshelled radius to the dry unshelled radius in order to have the dry and wet radius measurements - the wet unshelled radius must have been obtained before the sample cylinder was extracted.

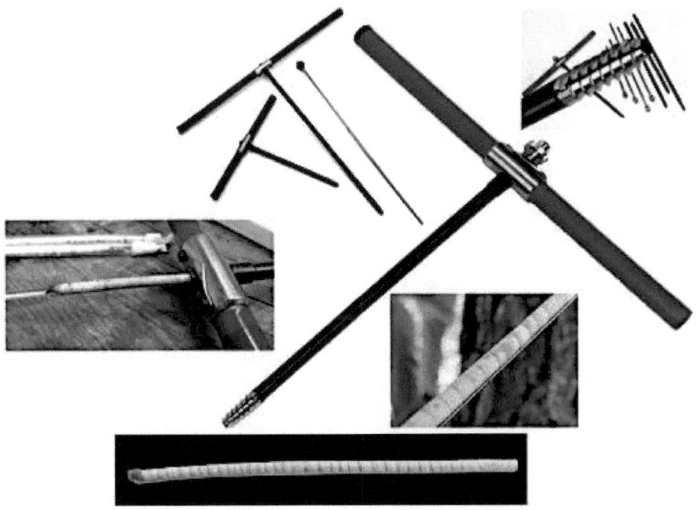

Figura 46 - Sampling for partial trunk analysis with auger.

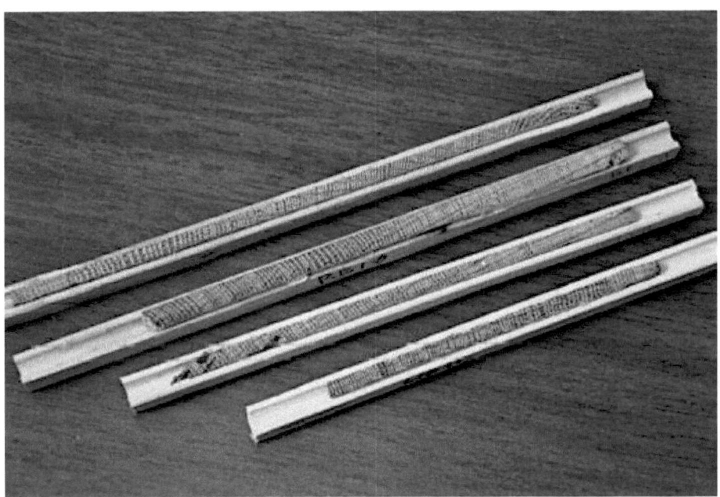

Figura 47 - Log sample cylinders obtained with an auger and glued to a wooden base for protection. Source: Theisen (2011).

8.10.2.2 Complete trunk analysis

Sample collection for complete trunk analysis in growth studies begins by selecting the trees for cubing using the same method recommended for cubing. The trees are felled and discs about 5 cm thick are removed as follows:

1) Note the geographical coordinates of the tree to be felled, identify the species, measure the diameter and thickness of the bark, as well as the total height of the tree;
2) A disk is removed from the base where the trunk was cut during felling;
3) Between 0.5 and 0.7 m high, a second disk is removed;
4) At chest level, at a height of 1.3 m, a third disk is cut;
5) Then a disk is extracted every meter, i.e. at a height of 2.3 m, 3.3 m, 4.3 m, and so on up to the top of the tree trunk;
6) If knots or other defects occur at the disc extraction point, the extraction position should be moved above the knot and the distance from the base to the disc cut position noted instead of the programmed position;
7) The discs should be given an identification plate with the tree number and the order number of the disc, numbered from 1 to n from the bottom to the top, preferably made of aluminum, which should be stapled to the more irregular side of the sample disc with a carpet stapler, as the regular side should be sanded after the discs have dried;
8) The tree should normally be cubed immediately after felling, measuring the diameter and thickness of the bark in the same positions where the discs for trunk analysis will be extracted;
9) Once the discs have been extracted and duly identified, they should be given an anti-fungal treatment and packed in raffia bags or other suitable packaging for transportation;
10) The discs must be properly spaced out in a suitable place to be dried in the air or in an oven;
11) After drying, the most even side, opposite the one that received the nameplate, should be sanded to allow the rings to be identified;

12)	Identify the largest radius between the pith and the beginning of the shell and mark this radius on the disk;

13)	The radii to be measured must also be marked in pencil on the disk and are 4 in number, the first being marked at 45° from the largest radius in a clockwise direction, the second at 90° from the first, the third at 90° from the second and the fourth radius at 90° from the third, as shown in Figure 48. Figura 48;

14)	Next, identify the position of each ring on each radius of the disk using a table magnifying glass and mark the position with a fine, soft pencil, which can be type B2 with a 0.5 mm pencil lead;

15)	Once the rings have been marked on all the radii of the disk, the measurement is made with the help of a laboratory magnifying glass, or the slices can be scanned and some image analysis or stem analysis software can be used.

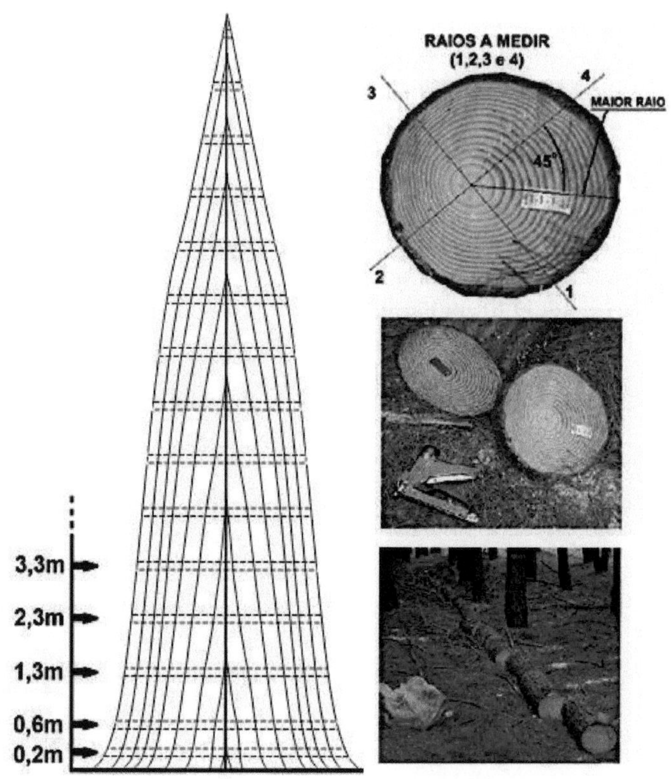

Figura 48 - Complete trunk analysis: disk sampling.

The profile graph of the trees at different ages can be produced in programs such as SAS, as in the following program:

```
* STEM ANALYSIS GRAPH - READ THE INSTRUCTIONS
BEGINNING WITH ASTERISKS TO USE/MODIFY THE
PROGRAM;
* THIS PROGRAM BUILDS GRAPHS OF THE TRUNK ANALYSIS
OF 12-YEAR-OLD TREES (RAYS 1 TO 12) - FOR A GREATER
NUMBER OF RAYS, IT IS NECESSARY TO INCLUDE IN THE
READING (INPUT) THE NUMBER (N) OF THE RAY (RAION)
```

AND THE COLUMNS THAT WILL BE READ UP TO THE AGE OF THE LAST RING, FOR EXAMPLE: RAIO13 47-49;
*---;
DATE A;
  ARRAY RADIUS{12} RADIUS1-RADIUS12;
  RETAIN RADIUS1-RADIUS12 0;
INPUT TREE 1-3 SECTION 4-5 RAION 6-7 RAIOCC 8-10 RAIO1 11-13 RAIO2 14-16
RADIUS3 17-19 RADIUS4 20-22 RADIUS5 23-25 RADIUS6 26-28 RADIUS7 29-31 RADIUS8 32-34
RADIUS9 35-37 RADIUS10 38-40 RADIUS11 41-43 RADIUS12 44-46;
* INPUT READS THE TREE NUMBER IN POSITIONS (COLUMNS) 1 TO 3, THE TREE SECTION NUMBER IN POSITIONS 4 TO 5, THE MEASURED RADIUS NUMBER IN POSITIONS 6 TO 7, THE OUTER RADIUS WITH BARK IN POSITIONS 8 TO 10 AND THEN, EVERY 3 POSITIONS, THE RADII OF EACH SLICE (RAIO1 TO RAION) - WHERE THERE IS NO MEASURED VALUE, A DOT MUST BE PLACED TO COMPLETE THE MAXIMUM NUMBER OF RADII PER SLICE UP TO THE TOTAL AGE OF THE TREE;
*---;
DO AGE=1 TO 12; RADIUS{AGE}=RADIUS{AGE}/10; END;
* THIS DO-END LOOP IDENTIFIES THE AGE OF EACH RAY - IF THERE ARE MORE THAN 12 RAYS, YOU NEED TO CHANGE THE MAXIMUM VALUE TO THE VALUE OF THE HIGHEST AGE - FOR EXAMPLE, FOR A MAXIMUM AGE OF 13, WRITE: DO AGE=1 TO 13;
*---;
  IF SECTION=1 THEN H=0.1;
  IF SECTION=2 THEN H=0.5;
  IF SECTION=3 THEN H=1.3;
  IF SECTION=4 THEN H=2.3;
  IF SECTION=5 THEN H=4.3;
  IF SECTION=6 THEN H=6.3;
  IF SECTION=7 THEN H=8.3;
  IF SECTION=8 THEN H=10.3;
  IF SECTION=9 THEN H=12.3;
  IF SECTION=10 THEN H=14.4;
  IF SECTION=11 THEN H=16.3;
  IF SECTION=12 THEN H=18.3;

    IF SECTION=13 THEN H=20.3;
* THE CONDITIONAL VALUE ASSIGNMENTS ABOVE REFER TO THE HEIGHT AT WHICH
  EACH SLICE HAS BEEN REMOVED FROM THE TREE - IF THERE ARE MORE SECTIONS, YOU SHOULD
  CREATE NEW DECLARATIONS - FOR EXAMPLE, FOR 14 SECTIONS WHERE IT WAS
  MEASURED AT A HEIGHT OF 22.3M, THE PROGRAM LINE MUST BE INCLUDED AT
  NEXT: IF SECTION=14 THEN H=22.3;
*---;
* THE DATA AFTER THE CARDS DECLARATION IS COPIED FROM THE ORIGINAL DATA FILES, INCLUDING THE NUMBER OF THE TREE IN THE FIRST 3 POSITIONS (COLUMNS) - SEE BELOW - YOU CAN INCLUDE AS MANY TREES AS YOU LIKE TO MAKE THE GRAPHS;
*---;
DATALINES;
001 1 1237217207201188176150133106 69 39 18 9
001 1 2198178170165155145126109 84 54 30 15 7
001 1 3194174169163150142123106 83 53 30 15 8
001 1 4230210203193184172145125100 66 40 20 9
001 2 1208191184177167157139120 91 64 37 23 9
001 2 2194176170166155144124107 85 51 29 14 7
001 2 3197179175170156143124108 86 55 31 15 8
001 2 4207189183175166154133113 90 60 35 15 8
001 3 1191175171166155145127112 89 57 35 15 .
001 3 2185169165160150138120104 81 50 28 12 .
001 3 3184168164159147137119105 83 54 29 13 .
001 3 4188172167159150138120102 82 55 34 12 .
001 4 1186173168157148138118101 82 53 28 9 .
001 4 2171158154148137126111 96 77 48 20 10 .
001 4 3173160156152140128111 96 76 49 27 9 .
001 4 4183170165160152142125109 86 55 29 9 .
001 5 1177164158153145136120101 78 48 20 . .
001 5 2158145139134125116102 88 67 38 17 . .
001 5 3163150147141129118102 85 63 35 15 . .
001 5 4176163158150140128110 94 73 43 19 . .
001 6 1159152146140129115 94 78 54 25 12 . .
001 6 2145138133128119108 92 74 51 24 12 . .
001 6 3157150145139128117 92 77 54 24 12 . .

```
001 6 4169162155146135121101 81 56 27 12 . .
001 7 1156152145137126110 84 58 34 7 . . .
001 7 2156152144135119103 79 55 34 9 . . .
001 7 3141138131124112 98 75 55 34 9 . . .
001 7 4139136131124114 99 76 55 31 9 . . .
001 8 1140128119113 94 80 52 31 12 . . . .
001 8 2131137128118104 85 55 34 12 . . . .
001 8 3127125118108 93 76 48 29 12 . . . .
001 8 4136133126114 99 83 53 32 12 . . . .
001 9 1117115104 89 72 55 30 9 . . . . .
001 9 2115113101 87 74 54 28 10 . . . . .
001 9 3114112101 88 73 56 29 9 . . . . .
001 9 4130128115 98 80 59 28 10 . . . . .
00110 1 73 71 59 41 21 12 . . . . . . .
00110 2 71 69 57 39 21 14 . . . . . . .
00110 3 69 68 56 39 22 13 . . . . . . .
00110 4 69 68 55 39 22 15 . . . . . . .
00111 1 49 48 44 33 19 . . . . . . . .
00111 2 46 45 41 33 17 . . . . . . . .
00111 3 43 42 39 28 16 . . . . . . . .
00111 4 43 42 38 29 18 . . . . . . . .
00112 1 12 11 7 3 . . . . . . . . .
00112 2 11 10 8 5 . . . . . . . . .
00112 3 12 11 9 6 . . . . . . . . .
00112 4 10 9 6 3 . . . . . . . . .
;
PROC SORT DATA=A; BY TREE SECTION RAION;
PROC SUMMARY DATA=A NWAY;
 CLASS TREE SECTION;
 VAR RADIUS1-RADIUS12 H;
 OUTPUT OUT=B MEAN(RADIUS1)=RADIUS1
MEAN(RADIUS2)=RADIUS2 MEAN(RADIUS3)=RADIUS3
        MEAN(RADIUS4)=RADIUS4
MEAN(RADIUS5)=RADIUS5 MEAN(RADIUS6)=RADIUS6
        MEAN(RADIUS7)=RADIUS7
MEAN(RADIUS8)=RADIUS8 MEAN(RADIUS9)=RADIUS9
        MEAN(RADIUS10)=RADIUS10
MEAN(RADIUS11)=RADIUS11 MEAN(RADIUS12)=RADIUS12
        MEAN(H)=H;
GOPTIONS VSIZE=10CM HSIZE=21CM HTEXT=5PCT
HTITLE=5PCT;
```

```
*------------------------------------------------------------------;
* AXIS1 IS THE AXIS OF THE ABSCISSES, WITH A MAXIMUM HEIGHT OF 20M - FOR HEIGHT
   LARGER CHANGE THE VALUES, FOR EXAMPLE FOR 30M: ORDER=(0 TO 30 BY 3);
AXIS1 ORDER=(0 TO 20 BY 2)
     LABEL=('HEIGHT (M)')
     MINOR=(N=3);
*------------------------------------------------------------------;
* AXIS2 IS THE AXIS OF ORDERS, WITH A MAXIMUM DIAMETER OF 21CM - IF THE DIAMETERS REACHED IN THE SLICES AT THE BASE OF THE TREE ARE LARGER, CHANGE THE VALUES, FOR EXAMPLE TO 40CM: ORDER=(0 TO 40 BY 4);
AXIS2 ORDER=(0 TO 21 BY 3)
     LABEL=(ANGLE=90 'RADIUS (CM)')
     MINOR=(N=2);
*------------------------------------------------------------------;
PROC SORT DATA=B; BY TREE SECTION;
PROC PRINT;
PROC GPLOT DATA=B;
  SYMBOL I=JOIN;
*------------------------------------------------------------------;
 * THE "SYMBOL I=JOIN" INSTRUCTION INSTRUCTS THE POINTS TO BE
   JOINED BY A LINE FORMING THE GRAPH;
*------------------------------------------------------------------;
 PLOT (RAIO1-RAIO12)*H / HREF=0 VREF=0 HAXIS=AXIS1 VAXIS=AXIS2 OVERLAY;
 BY ARVORE;
RUN;
```

The result printed by SAS is shown in Figura 49.

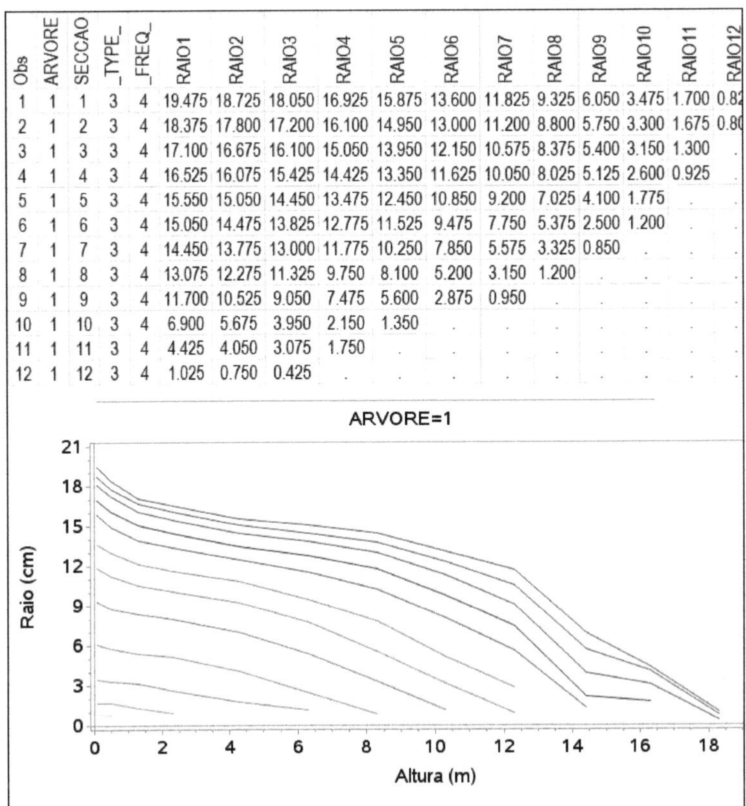

Obs	ARVORE	SECCAO	_TYPE_	_FREQ_	RAIO1	RAIO2	RAIO3	RAIO4	RAIO5	RAIO6	RAIO7	RAIO8	RAIO9	RAIO10	RAIO11	RAIO12
1	1	1	3	4	19.475	18.725	18.050	16.925	15.875	13.600	11.825	9.325	6.050	3.475	1.700	0.82
2	1	2	3	4	18.375	17.800	17.200	16.100	14.950	13.000	11.200	8.800	5.750	3.300	1.675	0.80
3	1	3	3	4	17.100	16.675	16.100	15.050	13.950	12.150	10.575	8.375	5.400	3.150	1.300	.
4	1	4	3	4	16.525	16.075	15.425	14.425	13.350	11.625	10.050	8.025	5.125	2.600	0.925	.
5	1	5	3	4	15.550	15.050	14.450	13.475	12.450	10.850	9.200	7.025	4.100	1.775	.	.
6	1	6	3	4	15.050	14.475	13.825	12.775	11.525	9.475	7.750	5.375	2.500	1.200	.	.
7	1	7	3	4	14.450	13.775	13.000	11.775	10.250	7.850	5.575	3.325	0.850	.	.	.
8	1	8	3	4	13.075	12.275	11.325	9.750	8.100	5.200	3.150	1.200	.	.	.	.
9	1	9	3	4	11.700	10.525	9.050	7.475	5.600	2.875	0.950	.	.	.	.	.
10	1	10	3	4	6.900	5.675	3.950	2.150	1.350	.	.	.	.	.	.	.
11	1	11	3	4	4.425	4.050	3.075	1.750	.	.	.	.	.	.	.	.
12	1	12	3	4	1.025	0.750	0.425	.	.	.	.	.	.	.	.	.

Figura 49 - Processing results for producing the trunk analysis graph, using SAS, for a 12-year-old tree.

9 DENDROCHRONOLOGY

Dendrochronology is the science of dating the growth rings of trees. Etymologically, the word comes from the Greek: dendron (tree) + chronos (time) + logos (study). It is a scientific method used in absolute dating applied to archaeology.

Dendrochronology consists of analyzing the annual growth ring patterns of trees to identify and date past occurrences, such as rockfall events, forest fires or snow avalanches, previous climatic conditions (ARX, 2021), silvicultural treatments, periods of excessive forest density, excessive pruning, pest or disease attacks, soil fertilization, etc.

Andrew Ellicott Douglass is considered the father of this science. He was an American astronomer, born on July 5, 1867, in Windsor, Vermont, and died on March 20, 1962, in Tucson, Arizona (LTRR, 2021). Douglass studied the cycles of sunspot activity and their relationship to weather patterns on Earth, which would result in tree growth ring patterns. This research led him to work on an archaeological project for the American Museum of Natural History, which included the dating of ruins by studying the growth ring patterns of the wood used in construction in relation to the tree ring patterns of the species used in the works that vegetated in the region. Later, at the University of Arizona, he was the first scientist to formally teach the subject of dendrochronology. In 1937, he founded the university's Tree Ring Research Laboratory, the first of its kind, consolidating the new science.

Trees grow in two directions, elongating and thickening. Elongation is promoted by the apical meristem, which comes from the seed embryo and is induced by gibberellins (hormones), causing growth in height. As the embryo elongates, the primary meristem differentiates just below the apex of the plant and forms the secondary meristem, which is induced by auxins, and which promotes the thickening of the trunk.

Trees that grow in regions with different seasons, dry and wet, or cold and hot, have their growth slowed down or even deactivated in the dry or cold seasons. In these cases, growth is said to be seasonal, i.e. it occurs during the time of year when there is enough moisture and heat to activate the plants' metabolism. At the beginning of the growing season, there is an abundance of hormones and nutrients, causing the secondary meristem to create large cells, with proportionally thinner cell walls compared to those generated when growth slows down, due to the less favorable environmental conditions at the end of the season. As a result, most plants form annual layers of stem cells, lighter at the beginning of the growing season and darker at the end. In a cross-section of the trunk, these layers are seen as light and dark rings, which can be counted and measured, allowing inferences to be made about the age of the tree and how much it has grown year on year.

In years of higher humidity and heat, or when the growing season is longer, trees form wider rings and vice versa. The patterns of wide and thin growth rings are like barcodes and make it possible to match them with the patterns of other trees, even after they have been transformed into beams and planks used in construction. It was this technique that Douglass used to date the time when certain buildings were constructed, creating dendrochronology.

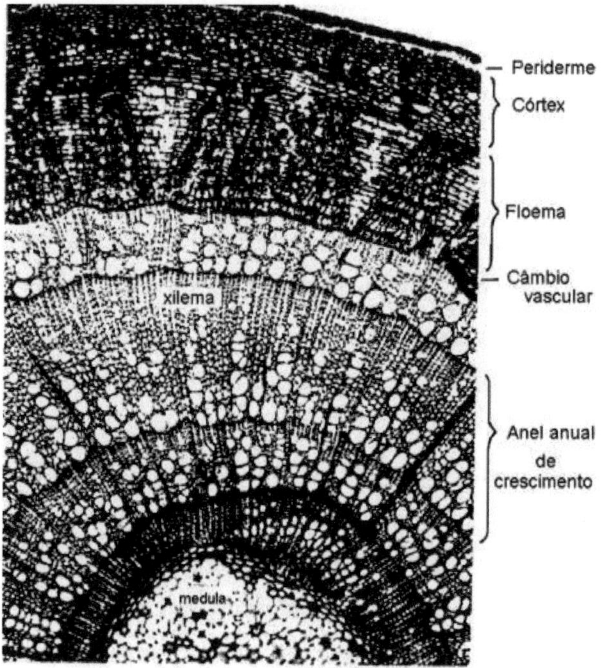

Figura 50 - Cross-section of a young stem of *Tilia olivieri*. Source: USP (2002).

Growth rings are also called tree rings or annual rings. The rings can form on trees that grow even in tropical regions without a dry season, due to the small climatic variations between the different times of the year.

The technique of comparing ring sequence patterns is called cross-dating and is considered the basis of dendrochronology. Various factors can influence the formation of rings, in addition to the climate, such as the conditions of the environment around each plant, as well as fires, or attacks by pests and diseases, solar flares, volcanic activity, glaciations, hurricanes, mass movements, earthquakes, among others, which can induce the formation of false rings and errors in the dating of rings. Trees in the same region tend to form similar ring width patterns. For this reason, correctly determining the age of a tree depends on cross-referencing information about its ring patterns with those of other trees for confirmation.

Dendrochronology has been used as an auxiliary science in archaeology, climate studies, environmental catastrophes of past eras and tree growth prognosis, the latter being the main application for forest production planning.

9.1 Cross-dating

Cross-dating is based on matching ring width patterns between different samples, allowing the date of ring formation to be defined, creating a time series. The technique is based on the fact that the width of annual rings tends to be a response to the environment. Thus, favorable years provide wider rings than years in which the environment was unfavorable for growth. A series of rings with different widths form patterns that make it possible to compare one sample with another and with the environmental conditions, making it possible to identify the year in which each growth ring was formed (Figura 51).

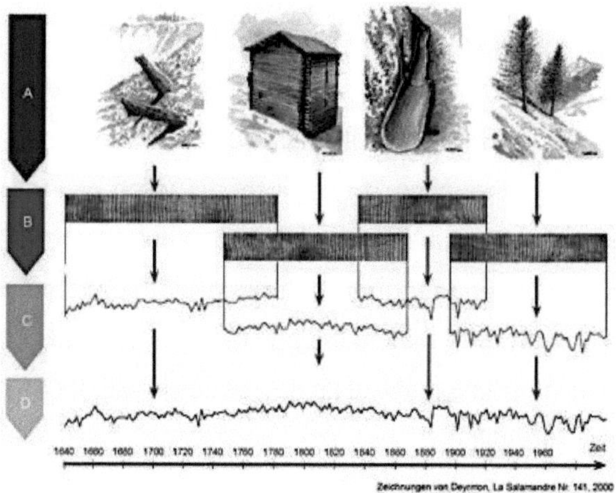

Figura 51 - General principle of cross-dating. (A) Search for living trees or wood at sites such as glacial deposits, old buildings, aqueducts; (B) recovery and preparation of wood samples; (C) measurement of the width and dating of annual growth rings; (D) determination of a chronology. Source: ETH (2021).

The crossing of sample ring patterns for cross-dating for archaeological purposes consists of using three types of samples: a sample of a living species, a sample of a dead species or wood fragment and a sample taken from archaeological contexts, such as old buildings. The crossing and overlapping of the three types of sample makes it possible to date archaeological samples and helps to calibrate regional standards in most cases (GONÇALVES, 2021).

After the measurement, the ring width patterns are synchronized to complement the cross-dating (Figura 52).

9.2 Replication

The rings of trees growing at the same time, in the same place, show small variations, and the arithmetic mean of the ring widths of trees from multiple samples must be calculated in order to construct a historical sequence of rings, which is known as replication.

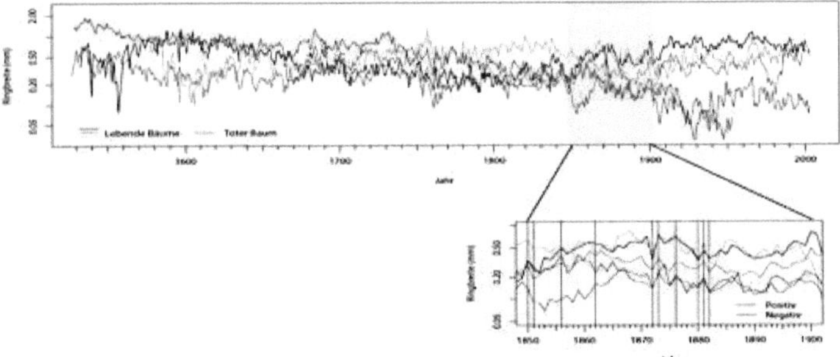

Figura 52 -Synchronized annual growth curves of Spruce (*Picea engelmannii*). Living trees are indicated as black lines, the dead tree is indicated as a red line, ring widths are shown on a logarithmic scale. The upper part of the figure shows the entire growth curves, the lower part focuses on the period 1850-1900 with visible positive and negative years being marked in blue and green respectively. Source: ETH (2021).

The sequence of tree rings whose start and end dates are not known is called a floating chronology. The floating chronology can be based on cross-referencing, combining the patterns of tree rings whose dates are known.

9.3 Radiocarbon dating

In addition to cross-dating and replication, the coherence between two independent sequences of annual rings can be confirmed by comparing their radiocarbon ages. Radiocarbon dating is a dating method using the unstable radioisotope carbon (C14) to date carbonaceous materials up to around 60,000 years old. C14 is found in the upper atmosphere where it combines with oxygen to form CO2 and descends into the biosphere where it is incorporated into plants by photosynthesis. The average lifespan of C14 is 5730 +/- 40 years. Counting the C14 atoms present in a dead organism makes it possible to know how many years have passed since it died, since its disintegration makes it possible to calculate how much time has passed since the organism's death. Radiocarbon dating must be calibrated to generate calendar dates, as C14 did not remain constant in the upper atmosphere. This technique was developed by Willard Libby and his colleagues at the University of Chicago in 1949.

10 FOREST BIOMASS

Biomass is the entire mass of organic matter of living and dead organisms, including parts of them in degradation or not, existing in a given community, above ground and integrated into the soil. Forest biomass is the tree component and phytomass of all plant components in the community, expressed as mass of dry matter (DM) per unit area (DM/ha) (HIGA, et al, 2014).

Therefore, the forest biomass in a stand is the sum of the mass of:

> Living trees;
> Necromass - dead trees and fallen branches with a diameter of more than 2 cm;
> Live understory plants ;
> Leaf litter, including dead animals and their remains;
> Soil organic matter.

Biomass studies can have different objectives, such as determining carbon stocks and availability for energy production. Depending on the objective, specific samples must be taken from each compartment to determine its carbon content or calorific value, which will not be dealt with in this text.

The carbon stored in the biomass in the various compartments represents around 47% of the dry mass of the total biomass.

The amount of biomass in the different compartments can be assessed by the direct method of destructive sampling, or by the indirect method using allometric equations adjusted for the same forest, or by equations adjusted for forests of the same type and characteristics.

PEARSON et al. (2005) report that carbon stock inventories can have a sampling error limit of 10%, i.e. 95% confidence probability for the mean.

Higa et al (2014) described the complete protocol adopted by EMBRAPA for assessing forest biomass in the book: "Protocolo de medição e estimativa de biomassa e carbono florestal" (Protocol for measuring and estimating forest biomass and carbon).

10.1 Live trees

The biomass of live trees in a forest stand is measured independently for the trunk (wood and bark), branches, leaves, flowers, fruit, seeds and roots.

The volume of the trunk is measured using traditional cubing methods. To transform it into biomass, it is necessary to measure and weigh representative samples along the trunk, previously dried, of the bark and wood, and determine their density.

The biomass of the branches is determined by weighing them while they are still green after felling the tree. Representative samples proportional to the size of the branches or the branching index are then taken, dried and weighed to determine their density, allowing the total mass of branches in the tree to be determined.

Leaf biomass is determined by weighing all the foliage immediately after felling. A representative sample is then taken and weighed green, dried and weighed again to determine its density and then determine the mass of all the leaves on the tree.

Sampling to determine the biomass of tree roots should be stratified by age class and size of the tree, history of disturbance and soil conditions (HIGA et al, 2014). It is generally carried out using trenches that should start as radii from the center of the trunk of the felled tree towards the end of the crown, and should be opened as far as the roots identified as belonging to the tree and greater than 2 mm in diameter can be found. The height, width and length of the trenches are measured to determine the volume of soil removed. The volume of soil containing all of the tree's roots is then calculated. The roots found are weighed green and then a representative sample is dried to determine its density, similar to the process used for branches. The mass of roots in the trench is then determined, which multiplied by the total volume of soil containing the tree's roots and divided by the volume of the trench, gives an estimate of the plant's root biomass.

The biomass of live trees in a stand is the sum of the biomass of all the trees growing in the stand. Therefore, the sample trees should include all diameter classes and the biomass determination should be calculated by the proportion of trees in each diameter class and then added together to find the biomass of the live trees in the stand.

10.2 Necromass

Necromass is represented by all fallen material on the ground with a diameter of 2 cm or more, not including dead plants that remain standing.

Necromass is generally sampled using linear transects, or any other type of sampling unit deemed appropriate.

The average diameters and lengths of all the fragments of trunks and branches in the sampling unit being used are measured.

When there are large numbers, representative samples can be taken from the fragments, which are measured, weighed wet and dried to determine their density. The fragments are separated into size classes to determine the mass per class and the total by adding up the mass of all the classes.

When they are small, they can be weighed as a whole while still in the field and samples taken to determine density.

10.3 Live understory plants

This includes lianas, epiphytes in general, tree seedlings, herbaceous and shrubby plants, etc. They should be sampled using units of a size appropriate to the size of the plants, and more than one size of sampling unit can be used. Preferably, the plants should be cut, weighed immediately after cutting, dried and weighed dry to determine their density, so that their mass can be calculated.

10.4 Mulch

The composition of litter varies according to the season due to the deposition of vegetation materials and the speed of decomposition and, according to HIGA et al (2014), litter studies must take into account the following factors: type of vegetation, latitude, altitude, temperature, rainfall, light availability during the growing season, day length (hours), evapotranspiration, relief, deciduousness, successional stage, herbivory, water availability and nutrient stocks in the soil.

10.5 Soil organic matter

Soil organic matter includes roots less than 2 mm in diameter. The sampling and determination of soil organic matter biomass must be carried out in accordance with soil sampling protocols, which differ from region to region and will not be dealt with in this text. Oliveira (2014) describes the protocol adopted by EMBRAPA as part of the PECUS research network in the book: "Protocolo para quantificação dos estoques de carbono do solo da rede de pesquisa Pecus".

10.6 Biomass and carbon stocks

According to the SFB (2020), the FAO recommends measuring carbon in each of the forest compartments listed below:

"Carbon in living biomass above ground - Includes trunks, branches, canopy, seeds and leaves.

Carbon in below-ground living biomass - Includes living roots, excluding those that are small (diameter < 2 millimeters) because they cannot be distinguished from soil organic matter or litter.

Carbon in dead biomass - Carbon in all dead woody biomass that is not part of the litter. It includes what has already fallen to the ground, dead roots and branches with diameters greater than 10 centimeters.

Carbon in litter - Carbon in all dead biomass with a diameter less than the minimum diameter required by Brazil to measure dead wood, in various stages of decomposition above the mineral or organic soil.

Soil carbon - Organic carbon in mineral and organic soils at a specific depth and applied consistently across all time series."

Biomass and carbon stocks in forests vary according to species, tree age, forest density, presence of understory, soil and climate.

Santos et al (2019) studied the carbon stock in the aerial part of a 19-year-old *Pinus elliottii* forest in a region with a latitude of 29° 43' and an altitude of 100m. The forest had an average diameter of 24.4 cm and a volume of 171.6 m³/ha, with a total aerial biomass stock of 173 t/ha and 74.7 t/ha of carbon (Tabela 8).

TABELA 8 - Biomass and carbon stock in the different components of *Pinus elliottii*

Compartment	Biomass		Carbon	
	t/ha	%	t/ha	%
Acicles	8,2	4,7	3,6	4,8
Twigs	14,8	8,5	5,5	7,3
Bark	16,5	9,5	6,2	8,3
Madeira	133,6	77,2	59,4	79,6
Total	173,0	100,0	74,7	100,0

Source: Santos et al (2029).

Gatto et al (2010) studied the carbon stock in eucalyptus stands cultivated in the central-eastern region of the state of Minas Gerais in different soil classes and found a total of 186.84 t/ha to 251.61 t/ha of carbon including above- and below-ground compartments up to a depth of 1 m (Tabela 9).

TABELA 9 - Total carbon stock (TCE) and partitioning in the soil-biomass system of eucalyptus plantations at 84 months of age

Compartimento	Região[1]					Média
	CO	RD	SA	SB	VI	
	ECT (t ha^{-1})					
Lenho	54,97	69,07	75,88	68,60	52,23	64,15
	(0,2 6)[2]	(0,37)	(0,30)	(0,31)	(0,23)	(0,29)
Resíduos[3]	30,55	36,98	40,19	36,58	30,17	34,89
	(0,14)	(0,20)	(0,16)	(0,17)	(0,13)	(0,16)
Solo (0–100 cm)	127,26	80,79	135,54	112,89	141,22	122,71
	(0,60)	(0,43)	(0,54)	(0,52)	(0,63)	(0,55)
Total	212,78	186,84	25 1,61	218,07	223,62	221,75
	(1,00)	(1,00)	(1,00)	(1,00)	(1,00)	(1,00)

[1] CO: Cocais; RD: Rio Doce; SA: Sabinópolis; SB: Santa Bárbara; VI: Virginópolis. [2] Partição do EC em relação ao ECT. [3] Resíduos da colheita (casca + folhas + galhos + raízes + manta orgânica).

Source: Gatto et al (2010).

In a study of stands between 2.3 and 8 years old of *Eucalyptus spp.* carried out by SILVA et al (2015), at a latitude of 23º to 23º 30', with average temperatures of 17.1ºC to 23.9ºC and annual rainfall of 1200 mm, 68.9% (27.11 kg/tree) of the carbon was found to be concentrated in the commercial part of the trunk and 31.1% (12.33 kg/tree) in the residues.

The SFB (2016) published the biomass stocks of Brazilian natural forests by biome in 2015, with a total for the country close to 117 billion tons (Tabela 10), with an average of 81.4% above ground, 17.3% below and 1.4% dead biomass.

TABELA 10 - Estimated amount of biomass in millions of metric tons for natural forests by biome, by compartment in 2015

Biome	Compartment	Millions of tons	%
Amazon	Above Ground Biomass	86.689	82,8%
	Below Ground Biomass	16.836	16,1%
	Dead Biomass	1.210	1,2%
Total Amazon		104.735	100,0%
Caatinga	Above Ground Biomass	2.014	76,2%
	Below Ground Biomass	544	20,6%
	Dead Biomass	84	0,1%
Total Caatinga		2.642	100,0%
Cerrado	Above Ground Biomass	3.613	62,6%
	Below Ground Biomass	2.117	36,7%
	Dead Biomass	42	0,0%
Total Cerrado		5.772	100,0%
Atlantic Forest	Above Ground Biomass	2.293	78,9%
	Below Ground Biomass	510	17,5%
	Dead Biomass	105	0,1%
Total Atlantic Forest		2.908	100,0%
Pampa	Above Ground Biomass	153	79,3%
	Below Ground Biomass	36	18,7%
	Dead Biomass	4	0,0%
Total Pampa		193	100,0%
Pantanal	Above Ground Biomass	482	66,9%
	Below Ground Biomass	232	32,2%
	Dead Biomass	6	0,0%
Total Pantanal		720	100,0%
Brazil	Above Ground Biomass	95.244	81,4%
	Below Ground Biomass	20.275	17,3%
	Dead Biomass	1.451	1,4%
Total Brazil		116.970	100,0%

Source: SFB (2016).

As you can see, the results vary greatly depending on the characteristics of the forest and the environment.

11 RELATIONSHIP BETWEEN DENDROMETRIC VARIABLES

Most of the dendrometric variables are highly correlated with each other, making it possible to estimate variables that are difficult to measure with those that are easier to measure. They are also correlated with time, making it possible to estimate their value as a function of age. The most common relationships between variables are as follows:

> Hypsometric relationship - this is the relationship between tree diameter (d) and height (h), allowing height to be estimated as a function of diameter using regression equations of the type h=f(d), known as hypsometric equations;

> Relation of diameter (d) and height (h) to volume (v) - allows the volume of a tree to be estimated as a function of diameter, or diameter and height using regression equations of the type v=f(d) and v=f(d, h), known as volumetric equations;

> Relationship of dendrometric variables (d, h, g, v) with time (t) - allows estimates to be made of any dendrometric variable (Y) over time (t) using growth equations of the type Y=f(t).

In mathematical models, the notation should be as follows:

> Y represents the dependent variable;

> X represents the independent variable;

- b_n represent the model coefficients for a sample;
- β_n represent the model parameters for a population.

It should be borne in mind that the relationships between the dependent (estimated) variables and the independent variables are generally not directly proportional and that the independent variables need to be transformed in order to obtain a relationship that adequately describes the relationship between the two. In the past, it was common to transform not only the independent variables but also the dependent variables, which is inappropriate and absolutely unnecessary, as the analysis of variance of the regression with the transformed variable is worthless and needs to be redone with the values without transformation. Another common mistake was to use models with 3, 4 or even more terms of the type $Y=b+b_{01}.X+b_2.X^2+b_3.X^3+b_4.X^4$. What happens is that equations with many terms are usually more distorted and become particular to the sample and cannot be generalized to the population from which the sample originated.

Therefore, the simplest models should be preferred and the independent variable should never be transformed.

Transformations of independent variables (X), among others, can be of the following types: X^2, X^3, X^4, X^5, $1/X$, $1/X^2$, $1/X^3$, $1/X^4$, $1/X^5$, $\ln(X)$, $\ln(X^2)$, $\ln(X^3)$, $\ln(X^4)$, $\ln(X^5)$, $\ln(1/X)$, $\ln(1/X^2)$, $\ln(1/X^3)$, $\ln(1/X^4)$, $\ln(1/X^5)$, $X^{1/2}$, $X^{1/3}$, $X^{1/4}$, $X^{1/5}$. Hypsometric relationship models should always be tested with and without an intercept, as often the equations generated without an intercept have a better fit.

11.1 Hypsometric equations

The hypsometric ratio is the relationship between tree height (h) and diameter (h) and can be represented by the function h = f (d). Some hypsometric relationship models are shown in Tabela 11.

TABELA 11 - Hypsometric relationship models

N⁰	Height Function Models
01	$h = d \, / \, (b_0 + b_1.d)$
02	$h = d(b_1.\ln d + b_2.\ln d^2)$
03	$h = b_0 + b_1.d^2$
04	$\square = b_0 + b_1.\ln d$
05	$\square = b_0 + b_1.d$
06	$h = b_0 + b_1.d + b_2.d^2$
07	$h = b_0 + b_1.\dfrac{1}{d^2}$
08	$\square = b_0 + b_1.\dfrac{1}{d} + b_2.\dfrac{1}{d^2}$
09	$\square = \left[\dfrac{b_0 + b_1.d + b_2.d^2}{d^2}\right] + 1{,}3$
10	$\square = b_0 + b_1.\ln\dfrac{1}{d}$
11	$\square = (b_0 + b_1.\ln d) + 1{,}3$

Where: d = Diameter at 1.30 m height, in centimeters; h = Tree height, in meters; b_0 , b_1 , b_2 , b_3 and b_4 = Equation coefficients; ln = Natural logarithm.

You can try to fit each of the models separately, but it is also possible to build a new model from all the possible transformations of the independent variable in software such as SAS, as in the following example, with data from the Tabela 17in

Appendix D, where the PROC REG procedure is run with the model including the intercept and excluding the intercept:

```
* HYPSOMETRIC RATIO;
DATA;
INPUT ARVORE D H @@;
* INDEPENDENT VARIABLE TRANSFORMATIONS;
D2=D**2; D3=D**3; D4=D**4; D5=D**5; ID=1/D; ID2=1/D2;
ID3=1/D3 ; ID4=1/D4 ; ID5=1/D5;
LD=LOG(D); LD2=LOG(D2); LD3=LOG(D3); LD4=LOG(D4);
LD5=LOG(D5);
LID=LOG(1/D); LID2=LOG(1/D2); LID3=LOG(1/D3);
LID4=LOG(1/D4); LID5=LOG(1/D5);
D12=D**(1/2); D13=D**(1/3); D14=D**(1/4); D15=D**(1/5);
DATALINES;
1 20.0 25.0 21 20.0 22.0 41 20.0 22.0
2 21.0 27.0 22 21.0 21.0 42 21.0 21.0
3 22.0 28.5 23 22.0 24.2 43 22.0 19.8
4 23.0 30.0 24 23.0 23.0 44 23.0 23.0
5 24.0 31.8 25 24.0 21.6 45 24.0 26.4
6 25.0 33.0 26 25.0 22.5 46 25.0 22.5
7 26.0 34.1 27 26.0 28.6 47 26.0 26.0
8 27.0 35.3 28 27.0 29.7 48 27.0 27.0
9 28.0 36.6 29 28.0 28.0 49 28.0 25.2
10 29.0 37.5 30 29.0 31.9 50 29.0 31.9
11 30.0 38.2 31 30.0 33.0 51 30.0 33.0
12 31.0 39.0 32 31.0 34.1 52 31.0 31.0
13 32.0 39.4 33 32.0 35.2 53 32.0 32.0
14 33.0 40.0 34 33.0 29.7 54 33.0 33.0
15 34.0 40.5 35 34.0 37.4 55 34.0 37.4
16 35.0 41.0 36 35.0 35.0 56 35.0 35.0
17 36.0 41.3 37 36.0 32.4 57 36.0 32.4
18 37.0 41.7 38 37.0 37.0 58 37.0 40.7
19 38.0 42.0 39 38.0 41.8 59 38.0 34.2
20 39.0 42.0 40 39.0 35.1 60 39.0 39.0
;
PROC REG DATA=DATA;
MODEL H = D D2 D3 D4 D5 ID ID2 ID3 ID4 ID5 LD LD2 LD3 LD4
LD5
     LID LID2 LID3 LID4 LID5 D12 D13 D14 D15 /
SELECTION=STEPWISE VIF;
```

```
RUN;
PROC REG DATA=DATA;
MODEL H = D D2 D3 D4 D5 ID ID2 ID3 ID4 ID5 LD LD2 LD3 LD4
LD5
        LID LID2 LID3 LID4 LID5 D12 D13 D14 D15 / NOINT
SELECTION=STEPWISE VIF;
RUN;
```

The SAS report with the results of the processing is listed below:

The REG Procedure
Model: MODEL1
Dependent Variable: h

Number of Observations Read	60
Number of Observations Used	60

Stepwise Selection: Step 1
Variable Ld3 Entered: R-Square = 0.6951 and C(p) = -1.9395

Analysis of Variance

Source	DF	Sum of Squares	Mean Square	F Value	Pr > F
Model	1	1760.33920	1760.33920	132.21	<.0001
Error	58	772.25813	13.31480		
Corrected Total	59	2532.59733			

Variable	Parameter Estimate	Standard Error	Type II SS	F Value	Pr > F
The Intercept	-58.68638	7.90046	734.68969	55.18	<.0001
Ld3	8.98372	0.78131	1760.33920	132.21	<.0001

Bounds on condition number: 1, 1
All variables left in the model are significant at the 0.1500 level.
No other variable met the 0.1500 significance level for entry into the model.

Summary of Stepwise Selection

Step	Variable Entered	Variable Removed	Number Vars In	Partial R-Square	Model R-Square	C(p)	F Value	Pr > F
1	Ld3		1	0.6951	0.6951	-1.9395	132.21	<.0001

The REG Procedure
Model: MODEL1
Dependent Variable: h

Number of Observations Read	60
Number of Observations Used	60

Analysis of Variance					
Source	DF	Sum of Squares	Mean Square	F Value	Pr > F
Model	1	1760.33920	1760.33920	132.21	<.0001
Error	58	772.25813	13.31480		
Corrected Total	59	2532.59733			

Root MSE		3.64894	R-Square	0.6951
Dependent Mean		31.99333	Adj R-Sq	0.6898
Coeff Var		11.40533		

Parameter Estimates						
Variable	DF	Parameter Estimate	Standard Error	t Value	Pr > |t|	Variance Inflation
The Intercept	1	-58.68638	7.90046	-7.43	<.0001	0
Ld3	1	8.98372	0.78131	11.50	<.0001	1.00000

As can be seen, in this model with intercept only the transformed variable Ld3, corresponding to $\ln(d^3)$ was retained in the equation, which showed a Coefficient of Determination of R^2 = 0.6951 and a Standard Error of Estimates (Syx%, also called Coefficient of Variation) of 11.41%. The graphs resulting from the processing are shown below.

The REG Procedure
Model: MODEL1
Dependent Variable: h

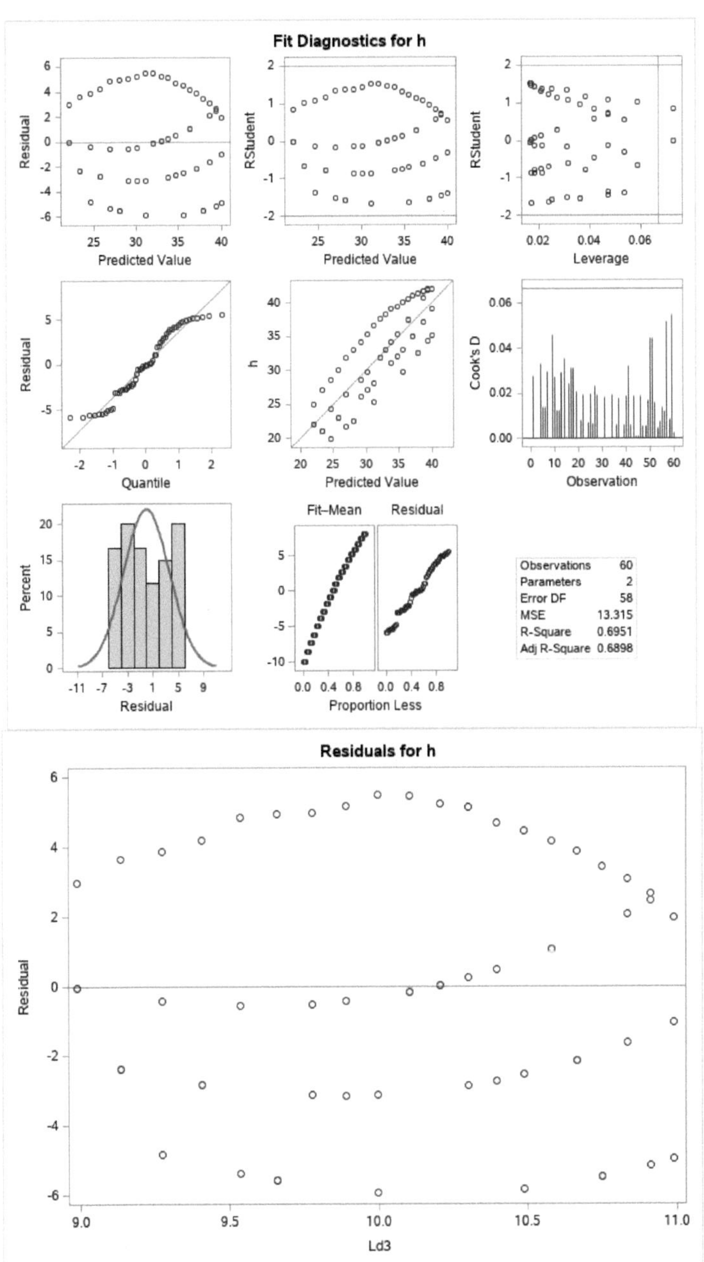

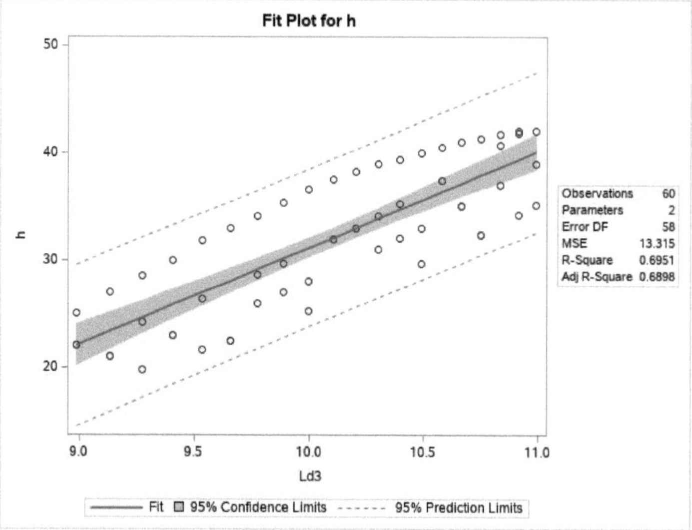

The REG Procedure
Model: MODEL1
Dependent Variable: h

Number of Observations Read	60
Number of Observations Used	60

Stepwise Selection: Step 1
Variable d Entered: R-Square = 0.9868 and C(p) = 1.1296
Note:No intercept in model. R-Square is redefined.

Analysis of Variance					
Source	DF	Sum of Squares	Mean Square	F Value	Pr > F
Model	1	63102	63102	4407.55	<.0001
Error	59	844.69491	14.31686		
Uncorrected Total	60	63947			

Variable	Parameter Estimate	Standard Error	Type II SS	F Value	Pr > F
d	1.07890	0.01625	63102	4407.55	<.0001

Bounds on condition number: 1, 1
Stepwise Selection: Step 2
Variable d5 Entered: R-Square = 0.9880 and C(p) = -2.0842
Note:No intercept in model. R-Square is redefined.

Analysis of Variance					
Source	DF	Sum of Squares	Mean Square	F Value	Pr > F
Model	2	63177	31588	2378.73	<.0001
Error	58	770.21315	13.27954		
Uncorrected Total	60	63947			

Variable	Parameter Estimate	Standard Error	Type II SS	F Value	Pr > F
d	1.14408	0.03166	17340	1305.77	<.0001
d5	-5.54087E-8	2.339615E-8	74.48176	5.61	0.0212

Bounds on condition number: 4.0921, 16.368
All variables left in the model are significant at the 0.1500 level.
No other variable met the 0.1500 significance level for entry into the model.
Note:No intercept in model. R-Square is redefined.

Summary of Stepwise Selection								
Step	Variable Entered	Variable Removed	Number Vars In	Partial R-Square	Model R-Square	C(p)	F Value	Pr > F
1	d		1	0.9868	0.9868	1.1296	4407.55	<.0001
2	d5		2	0.0012	0.9880	-2.0842	5.61	0.0212

The REG Procedure
Model: MODEL1
Dependent Variable: h

Number of Observations Read	60
Number of Observations Used	60

Note:No intercept in model. R-Square is redefined.

Analysis of Variance					
Source	DF	Sum of Squares	Mean Square	F Value	Pr > F
Model	2	63177	31588	2378.73	<.0001
Error	58	770.21315	13.27954		
Uncorrected Total	60	63947			

Root MSE	3.64411	R-Square	0.9880
Dependent Mean	31.99333	Adj R-Sq	0.9875
Coeff Var	11.39022		

Parameter Estimates						
Variable	DF	Parameter Estimate	Standard Error	t Value	Pr > |t|	Variance Inflation
d	1	1.14408	0.03166	36.14	<.0001	4.09208
d5	1	-5.54087E-8	2.339615E-8	-2.37	0.0212	4.09208

The results of the modeling without an intercept were better and retained the variables d and d5, the latter corresponding to d^5 , with the best R^2 of 0.9880 and a similar Syx of 11.39%. The analysis graphs are on the next page. The Variance Inflation Index was 4.09 for the two variables retained in the equation, which is less than 10, making it unnecessary to exclude variables from the equation.

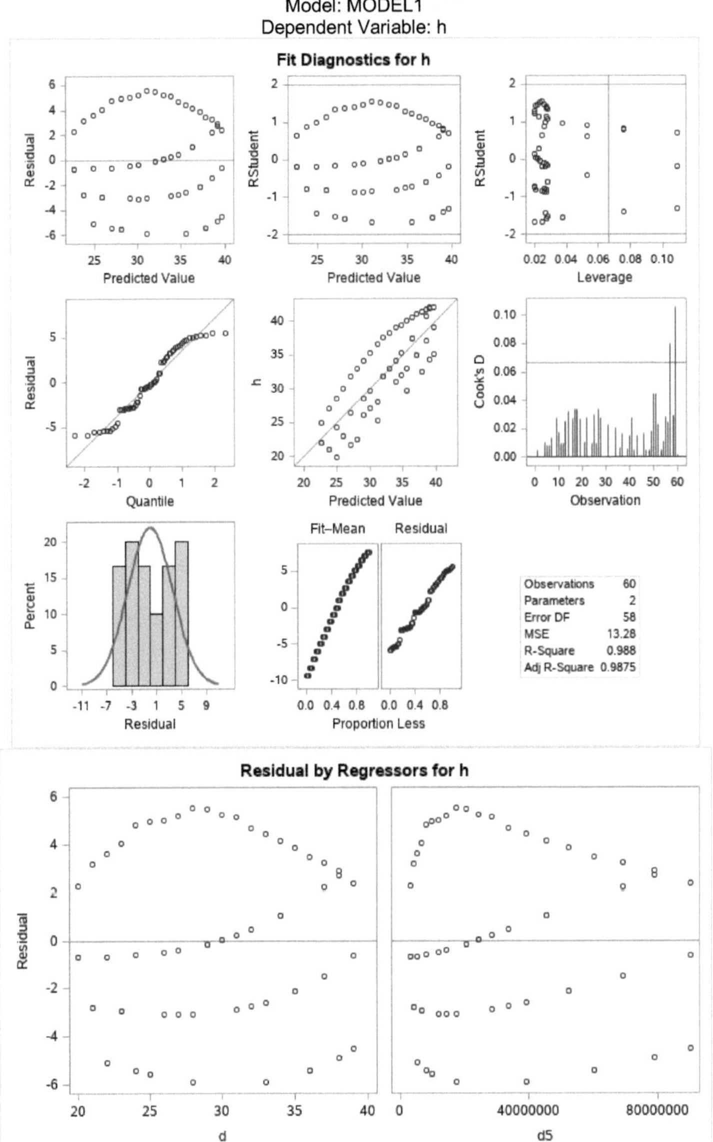
Fit Diagnostics for h
Residual
Predicted Value
RStudent
Predicted Value
RStudent
Leverage
Residual
Quantile
h
Predicted Value
Cook's D
Observation
Percent
Residual
Fit–Mean
Residual
Proportion Less
Observations 60
Parameters 2
Error DF 58
MSE 13.28
R-Square 0.988
Adj R-Square 0.9875
Residual by Regressors for h
Residual
d
d5

11.2 Volumetric equations

Many mathematical models have been developed for volumetric estimates based on diameter and height ($v = f(d, h)$). A study with an excessive number of models was carried out by Silvestre et al (2014), who adjusted 33 models for volumetric estimates of *Pinus taeda* L. stands in the municipality of Lages, SC, as shown in Tabela 12.

11.3 Growth equations

Growth can be modeled using empirical or phenomenological equations, for example:

> ➢ Empirical: equations in which the parameters have no relation to the biological phenomena of growth; example: polynomials in general;
>
> ➢ Phenomenological: equations in which the parameters correspond to characteristics of the growth curve; example: Chapman-Richards.

Two models that fit most tree growth curves are the 5th degree polynomial and the Chapman-Richards equation, especially for data that has a sigmoid-shaped point cloud on a graph.

TABELA 12 - **Mathematical models used to adjust volumetric equations by Silvestre et al (2014).**

N°	Autor	Modelo
1	Burkhault I	$v = \beta 0 + DAP\beta 1 + ei$
2	Kopezky-Gehrhatdt I	$v = \beta 0 + \beta 1 \cdot DAP + ei$
3	Kopezky-Gehrhatdt II	$v = \beta 0 + \beta 1 \cdot DAP^2 + ei$
4	Dissescu-Meyer	$v = \beta 1 \cdot DAP + \beta 2 \cdot DAP^2 + ei$
5	HohenadlKrenn	$v = \beta 0 + \beta 1 \cdot DAP + \beta 2 \cdot DAP^2 + ei$
6	Bonetes I	$v = \beta 0 + \beta 1 \cdot (1/DAP) + \beta 2 \cdot (1/DAP)^2 + ei$
7	Husch-1963	$Log\,(v) = \beta 0 + \beta 1 \cdot log\,(DAP) + ei$
8	Omerod MI	$v = \beta 0 + \beta 1 \cdot DAP \cdot h^2 + \beta 2 \cdot DAP^2 + ei$
9	Omerod MII	$v = \beta 0 + \beta 1 \cdot DAP^2 \cdot h + \beta 2 \cdot (1/\sqrt{DAP}) + ei$
10	Spurr FF constante	$v = \beta 1 \cdot DAP^2 \cdot h + ei$
11	Spurr Var combinada	$v = \beta 0 + \beta 1 \cdot DAP \cdot h + ei$
12	Spurr Var combinada II	$v = \beta 0 + \beta 1 \cdot DAP^2 \cdot h + ei$
13	Bonetes II	$v = \beta 0 + \beta 1 \cdot DAP + \beta 2 \cdot h + ei$
14	Bonetes III	$v = \beta 0 + \beta 1 \cdot DAP^2 + \beta 2 \cdot h + ei$
15	Bonetes IV	$v = \beta 0 + \beta 1 \cdot DAP^2 + \beta 2 \cdot h^2 + ei$
16	Stoate australiana	$v = \beta 0 + \beta 1 \cdot DAP^2 + \beta 2 \cdot DAP^2 \cdot h + \beta 3 \cdot h + ei$
17	Burkhaut II	$v = \beta 0 + \beta 1 \cdot \sqrt{DAP} \cdot h + \beta 2 \cdot (1/DAP) + \beta 3 \cdot DAP^2 + ei$
18	Burkhaut III	$v = \beta 0 + \beta 1 \cdot h^2 + \beta 2 \cdot DAP^2 + \beta 3 \cdot (1/\sqrt{DAP}) + ei$
19	Meyer modificada I	$v = \beta 0 + \beta 1 \cdot DAP \cdot h + \beta 2 \cdot DAP^2 \cdot h^2 + \beta 3 \cdot DAP^3 \cdot h^3 + \beta 4 \cdot DAP4 \cdot h4 + ei$
20	Meyer modificada II	$v = \beta 0 + \beta 1 \cdot DAP + \beta 2 \cdot DAP^2 + \beta 3 \cdot DAP \cdot h + \beta 4 \cdot DAP^2 \cdot h + ei$
21	Meyer compreensiva	$v = \beta 0 + \beta 1 \cdot DAP + \beta 2 \cdot DAP^2 + \beta 3 \cdot DAP \cdot h + \beta 4 \cdot DAP^2 \cdot h + \beta 5 \cdot h + ei$
22	Näslund-spruce I	$v = \beta 0 + \beta 1 \cdot DAP^2 + \beta 2 \cdot DAP^2 \cdot h + \beta 3 \cdot DAP \cdot h^2 + \beta 4 \cdot h + ei$
23	Naslund-spruce II	$v = \beta 0 + \beta 1 \cdot DAP^2 + \beta 2 \cdot ln\,(DAP^2 \cdot h) + \beta 3 \cdot DAP \cdot h^2 + \beta 4 \cdot h^2 + ei$
24	Clutter	$v = e\beta 0 + \beta 1 \cdot h + \beta 2 \cdot log\,(DAP) + ei$
25	Clutter II	$v = e\beta 0 + \beta 1 \cdot log^2\,(h) + \beta 2 \cdot log\,(DAP) + ei$
26	Takata	$v = DAP \cdot h/(\beta 0 + \beta 1 \cdot (1/DAP)) + ei$
27	Spurr-Var combinada III	$ln\,(v) = \beta 0 + \beta 1 \cdot ln\,(DAP^2 \cdot h) + ei$
28	I.B.W. Alemanha	$ln\,(v) = \beta 0 + \beta 1 \cdot ln\,(DAP) + \beta 2 \cdot ln^2\,(DAP) + \beta 3 \cdot ln\,(h) + \beta 4 \cdot ln^2\,(h) + ei$
29	Schumacher Hall não linear	$v = \beta 0 \cdot (DAP\beta 1) \cdot (h\beta 2) + ei$
30	Schumacher Hall	$ln\,(v) = \beta 0 + \beta 1 \cdot ln\,(DAP) + \beta 2 \cdot ln\,(h) + ei$
31	Stoate	$Log\,(v) = \beta 0 + \beta 1 \cdot log^2\,(h) + \beta 2 \cdot log^2\,(DAP) + \beta 3 \cdot DAP^2 + ei$
32	Prodan II	$Log(v) = \beta 0 + \beta 1 \cdot log\,(d) + \beta 2 \cdot log^2\,(DAP) + \beta 3 \cdot log\,(h) + \beta 4 \cdot log^2\,(h) + ei$
33	Prodan I	$\sqrt{v} = \beta 0 + \beta 1 \cdot DAP \cdot h + \beta 2 \cdot DAP^2 \cdot h^2 + \beta 3 \cdot DAP^3/h^3 + \beta 4 \cdot DAP4/h4 + \beta 5 \cdot DAP5 \cdot h5 + \beta 6 \cdot DAP6 \cdot h6 + ei$

Em que: v = volume; DAP = diâmetro à altura do peito (1,30 m); h = altura total; $\beta 0, \beta 1, \beta 2, \beta 3, \beta 4, \beta 5, \beta 6$ = coeficientes; ln = logaritmo neperiano; e = exponencial; log = logaritmo base 10; ei = erro aleatório.

The 5th degree polynomial is expressed as:

$$Y = b_0 + b_1\,t + b_2\,t^2 + b_3\,t^3 + b_4\,t^4 + b_5\,t^5$$

The Chapman-Richards equation is expressed by (Sit, 1994):

$$Y = A \cdot (1 - e)^{-k.tr}$$

where: Y = value of the dependent dendrometric variable at age t; b_0, b_1, b_2, b_3, b_4, b_5 = parameters of the polynomial; A = parameter representing the upper asymptote (>0); k = parameter representing the speed of growth (>0 and <1); r = parameter representing the inflection point of the curve (>0); t = independent variable (age).

In Tabela 13 shows data from an example of fitting the two growth models for six *Pinus elliottii* trees at 25 years of age.

The first step in modeling is to draw graphs of the variables on the ordinate in relation to age on the abscissa (Figura 53). Then, the shape of the data is checked in order to select equations that fit the data. Sit (1994) and Kiviste et al (2002) list various growth models with their characteristics and the respective graphs they can take, making it easier to choose the models to test.

It can be seen that the diameter, height and basal area data take on approximately sigmoid shapes, while the volume data is typically exponential (Figura 53). Therefore, for the first three variables you can use models that have an inflection point and a horizontal asymptote of maximum value. For volume, it is necessary to look for an equation that can take the exponential form. What can be seen in relation to volume is that, at the older age of the measurement, it seems to be close to the inflection point of the curve, due to the growth trend in recent years.

TABELA 13 - Growth data of 6 Pinus elliottii at 25 years of age

n	t	d	h	v	n	t	d	h	v	n	t	d	h	v
1	25	25,8	27,7	0,795	2	25	28	26,7	0,905	3	25	21,1	23,7	0,45
1	24	25,1	26,9	0,729	2	24	27,6	26,2	0,847	3	24	20,6	23,1	0,419
1	23	24,3	26,1	0,674	2	23	26,9	25,7	0,783	3	23	20,1	22,5	0,387
1	22	23,6	25	0,614	2	22	26,3	25,1	0,722	3	22	19,6	21,9	0,358
1	21	22,9	23,8	0,55	2	21	25,7	24,4	0,654	3	21	19,1	21,3	0,326
1	20	22,1	22,8	0,491	2	20	24,9	23,5	0,589	3	20	18,6	20,5	0,293
1	19	21,3	21,8	0,43	2	19	24,2	22,4	0,525	3	19	18	19,6	0,262
1	18	20,4	20,8	0,373	2	18	23,5	21,4	0,474	3	18	17,5	18,6	0,237
1	17	19,5	19,7	0,324	2	17	22,8	20,4	0,421	3	17	17	17,5	0,212
1	16	18,5	18,7	0,276	2	16	22,3	19,4	0,372	3	16	16,5	16,5	0,184
1	15	17,7	17,7	0,238	2	15	21,8	18,4	0,338	3	15	16,1	15,6	0,169

n	t	d	h	v	n	t	d	h	v	n	t	d	h	v
1	14	16,8	16,7	0,2	2	14	20,8	17	0,288	3	14	15,6	14,5	0,149
1	13	15,8	15,5	0,163	2	13	19,8	15,6	0,241	3	13	15	13,3	0,127
1	12	14,6	14,5	0,129	2	12	18,8	14,6	0,196	3	12	14,3	12,1	0,103
1	11	13,5	13,4	0,101	2	11	17,9	13,5	0,163	3	11	13,5	10,8	0,083
1	10	12,8	12,1	0,08	2	10	16,8	12,2	0,132	3	10	12,6	9,5	0,064
1	9	11,7	10,8	0,06	2	9	15,7	10,8	0,105	3	9	11,5	8,4	0,048
1	8	10,4	9,1	0,041	2	8	14,3	9,6	0,078	3	8	10,2	7,2	0,033
1	7	8,1	7	0,023	2	7	12,3	8,5	0,052	3	7	8,6	6	0,021
1	6	6,1	5,3	0,011	2	6	10,2	7	0,032	3	6	6,9	4,7	0,011
1	5	4,5	4	0,004	2	5	8,2	5,4	0,016	3	5	5,1	3,5	0,005
1	4	2,8	2,4	0,002	2	4	6,6	4,1	0,009	3	4	3,7	2,5	0,003
1	3	0	0,9	0	2	3	4,2	2,6	0,003	3	3	1,7	1,5	0,001
1	2	0	0,5	0	2	2	0,8	1	0	3	2	0	0,7	0
1	1	0	0,3	0	2	1	0	0,3	0	3	1	0	0,5	0

Keep going ...

n	t	d	h	v	n	t	d	h	v	n	t	d	h	v
4	25	26,6	24,6	0,754	5	25	32,2	30,6	1,336	6	25	27,6	26,4	0,77
4	24	26,2	23,9	0,714	5	24	31,6	29,2	1,262	6	24	27,1	24,8	0,724
4	23	25,7	23,2	0,67	5	23	30,9	27,9	1,185	6	23	26,5	23,8	0,679
4	22	25,1	22,4	0,623	5	22	30,3	26,9	1,107	6	22	26,1	23,2	0,634
4	21	24,5	21,5	0,565	5	21	29,7	25,8	1,013	6	21	25,5	22,5	0,58
4	20	23,8	20,6	0,508	5	20	29	24,8	0,916	6	20	24,9	21,7	0,528
4	19	23	19,7	0,452	5	19	28,4	23,2	0,836	6	19	24,2	20,7	0,477
4	18	22,2	18,7	0,402	5	18	27,5	21,9	0,75	6	18	23,5	19,8	0,424
4	17	21,4	17,6	0,348	5	17	26,9	21,2	0,678	6	17	23,1	19	0,387
4	16	20,5	16,6	0,292	5	16	26,1	20,6	0,594	6	16	22,5	18,3	0,343
4	15	20	15,6	0,265	5	15	25,6	19,7	0,549	6	15	22	17,5	0,311
4	14	19,3	14,6	0,229	5	14	24,7	18,7	0,475	6	14	21,2	16,3	0,265
4	13	18,4	13,6	0,197	5	13	23,5	17,5	0,4	6	13	20,5	14,9	0,222
4	12	17,3	12,5	0,161	5	12	22	16,4	0,315	6	12	19,6	13,1	0,177
4	11	16,3	11,5	0,132	5	11	20,6	14,9	0,244	6	11	18,8	11,6	0,143
4	10	15	10,5	0,103	5	10	19,1	13,1	0,185	6	10	18,1	10,6	0,116
4	9	13,7	9,3	0,077	5	9	17	11,5	0,131	6	9	16,8	9,4	0,09
4	8	11,7	7,9	0,051	5	8	15,2	10,3	0,09	6	8	15,3	8,1	0,063
4	7	9,7	6,6	0,03	5	7	13,2	8,8	0,057	6	7	13,5	6,7	0,041
4	6	7,9	5,3	0,017	5	6	11,1	6	0,034	6	6	11,5	5	0,026
4	5	6	4,1	0,008	5	5	8,9	4	0,019	6	5	9	3,4	0,014
4	4	4,3	2,8	0,004	5	4	6,5	3,5	0,01	6	4	6,6	2,4	0,007
4	3	2,4	1,6	0,001	5	3	3,4	2,2	0,003	6	3	3,9	1,6	0,003
4	2	0	0,9	0	5	2	0	0,9	0	6	2	0	0,9	0
4	1	0	0,5	0	5	1	0	0,5	0	6	1	0	0,4	0

Where: n = tree number; t = age (years); d = diameter (cm); h = height (m); v = volume (m³).

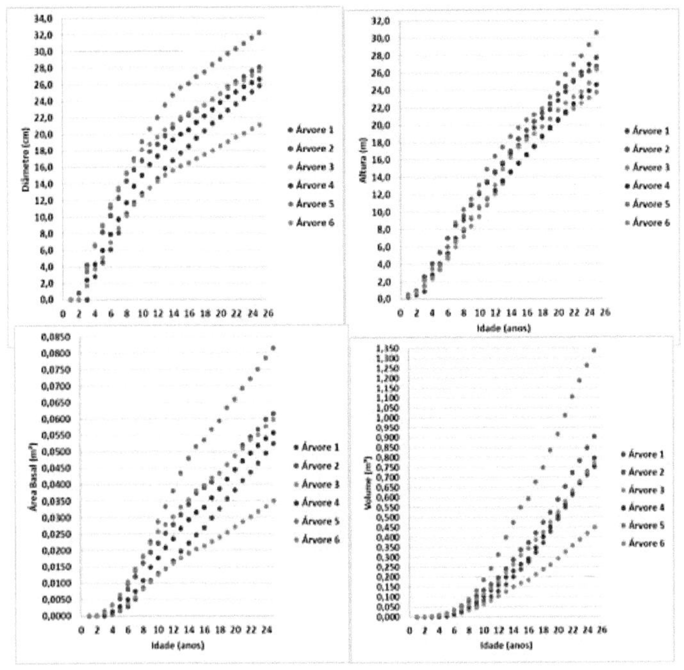

Figura 53 - Growth chart of 6 Pinus elliottii at 25 years of age.

The equations were adjusted using SAS® OnDemand for Academics with the following program:

```
* GROWING 6 PINE TREES;
DATA DATA;
INPUT N T D H V @@;
G=3.141593*D**2/40000;
T2=T**2; T3=T**3; T4=T**4; T5=T**5;
DATALINES;
1 25 25.8 27.7 0.795 2 25 28.0 26.7 0.905 3 25 21.1 23.7 0.450 4
25 26.6 24.6 0.754 5 25 32.2 30.6 1.336 6 25 27.6 26.4 0.770
1 24 25.1 26.9 0.729 2 24 27.6 26.2 0.847 3 24 20.6 23.1 0.419 4
24 26.2 23.9 0.714 5 24 31.6 29.2 1.262 6 24 27.1 24.8 0.724
1 23 24.3 26.1 0.674 2 23 26.9 25.7 0.783 3 23 20.1 22.5 0.387 4
23 25.7 23.2 0.670 5 23 30.9 27.9 1.185 6 23 26.5 23.8 0.679
1 22 23.6 25.0 0.614 2 22 26.3 25.1 0.722 3 22 19.6 21.9 0.358 4
```

22 25.1 22.4 0.623 5 22 30.3 26.9 1.107 6 22 26.1 23.2 0.634
1 21 22.9 23.8 0.550 2 21 25.7 24.4 0.654 3 21 19.1 21.3 0.326 4
21 24.5 21.5 0.565 5 21 29.7 25.8 1.013 6 21 25.5 22.5 0.580
1 20 22.1 22.8 0.491 2 20 24.9 23.5 0.589 3 20 18.6 20.5 0.293 4
20 23.8 20.6 0.508 5 20 29.0 24.8 0.916 6 20 24.9 21.7 0.528
1 19 21.3 21.8 0.430 2 19 24.2 22.4 0.525 3 19 18.0 19.6 0.262 4
19 23.0 19.7 0.452 5 19 28.4 23.2 0.836 6 19 24.2 20.7 0.477
1 18 20.4 20.8 0.373 2 18 23.5 21.4 0.474 3 18 17.5 18.6 0.237 4
18 22.2 18.7 0.402 5 18 27.5 21.9 0.750 6 18 23.5 19.8 0.424
1 17 19.5 19.7 0.324 2 17 22.8 20.4 0.421 3 17 17.0 17.5 0.212 4
17 21.4 17.6 0.348 5 17 26.9 21.2 0.678 6 17 23.1 19.0 0.387
1 16 18.5 18.7 0.276 2 16 22.3 19.4 0.372 3 16 16.5 16.5 0.184 4
16 20.5 16.6 0.292 5 16 26.1 20.6 0.594 6 16 22.5 18.3 0.343
1 15 17.7 17.7 0.238 2 15 21.8 18.4 0.338 3 15 16.1 15.6 0.169 4
15 20.0 15.6 0.265 5 15 25.6 19.7 0.549 6 15 22.0 17.5 0.311
1 14 16.8 16.7 0.200 2 14 20.8 17.0 0.288 3 14 15.6 14.5 0.149 4
14 19.3 14.6 0.229 5 14 24.7 18.7 0.475 6 14 21.2 16.3 0.265
1 13 15.8 15.5 0.163 2 13 19.8 15.6 0.241 3 13 15.0 13.3 0.127 4
13 18.4 13.6 0.197 5 13 23.5 17.5 0.400 6 13 20.5 14.9 0.222
1 12 14.6 14.5 0.129 2 12 18.8 14.6 0.196 3 12 14.3 12.1 0.103 4
12 17.3 12.5 0.161 5 12 22.0 16.4 0.315 6 12 19.6 13.1 0.177
1 11 13.5 13.4 0.101 2 11 17.9 13.5 0.163 3 11 13.5 10.8 0.083 4
11 16.3 11.5 0.132 5 11 20.6 14.9 0.244 6 11 18.8 11.6 0.143
1 10 12.8 12.1 0.080 2 10 16.8 12.2 0.132 3 10 12.6 9.5 0.064 4
10 15.0 10.5 0.103 5 10 19.1 13.1 0.185 6 10 18.1 10.6 0.116
1 9 11.7 10.8 0.060 2 9 15.7 10.8 0.105 3 9 11.5 8.4 0.048 4 9
13.7 9.3 0.077 5 9 17.0 11.5 0.131 6 9 16.8 9.4 0.090
1 8 10.4 9.1 0.041 2 8 14.3 9.6 0.078 3 8 10.2 7.2 0.033 4 8 11.7
7.9 0.051 5 8 15.2 10.3 0.090 6 8 15.3 8.1 0.063
1 7 8.1 7.0 0.023 2 7 12.3 8.5 0.052 3 7 8.6 6.0 0.021 4 7 9.7 6.6
0.030 5 7 13.2 8.8 0.057 6 7 13.5 6.7 0.041
1 6 6.1 5.3 0.011 2 6 10.2 7.0 0.032 3 6 6.9 4.7 0.011 4 6 7.9 5.3
0.017 5 6 11.1 6.0 0.034 6 6 11.5 5.0 0.026
1 5 4.5 4.0 0.004 2 5 8.2 5.4 0.016 3 5 5.1 3.5 0.005 4 5 6.0 4.1
0.008 5 5 8.9 4.0 0.019 6 5 9.0 3.4 0.014
1 4 2.8 2.4 0.002 2 4 6.6 4.1 0.009 3 4 3.7 2.5 0.003 4 4 4.3 2.8
0.004 5 4 6.5 3.5 0.010 6 4 6.6 2.4 0.007
1 3 0.0 0.9 0.000 2 3 4.2 2.6 0.003 3 3 1.7 1.5 0.001 4 3 2.4 1.6
0.001 5 3 3.4 2.2 0.003 6 3 3.9 1.6 0.003
1 2 0.0 0.5 0.000 2 2 0.8 1.0 0.000 3 2 0.0 0.7 0.000 4 2 0.0 0.9
0.000 5 2 0.0 0.9 0.000 6 2 0.0 0.9 0.000

```
1 1 0.0 0.3 0.000 2 1 0.0 0.3 0.000 3 1 0.0 0.5 0.000 4 1 0.0 0.5
0.000 5 1 0.0 0.5 0.000 6 1 0.0 0.4 0.000
;
PROC REG DATA=DATA;
   MODEL D=T T2 T3 T4 T5;
RUN;
PROC REG DATA=DATA;
   MODEL H=T T2 T3 T4 T5;
RUN;
PROC REG DATA=DATA;
   MODEL G=T T2 T3 T4 T5;
RUN;
PROC REG DATA=DATA;
   MODEL V=T T2 T3 T4 T5;
RUN;
PROC MODEL DATA=DATA;
* D = A*(1-EXP(-K*T))**R;
PARAMETERS A=40.0 K=0.05 R=2;
EKT = EXP(-K*T);
EKT1 = 1 - EKT;
EKTR = (EKT1)**R;
 D = A*EKTR;
 FIT D;
DER.A = EKTR;
DER.K = A*T*R*EKT*EKT1**(R-1);
DER.R = A*EKTR*LOG(EKT1);
RUN;
PROC MODEL DATA=DATA;
* H = A*(1-EXP(-K*T))**R;
PARAMETERS A=40.0 K=0.05 R=2;
EKT = EXP(-K*T);
EKT1 = 1 - EKT;
EKTR = (EKT1)**R;
 H = A*EKTR;
 FIT H;
DER.A = EKTR;
DER.K = A*T*R*EKT*EKT1**(R-1);
DER.R = A*EKTR*LOG(EKT1);
RUN;
PROC MODEL DATA=DATA MAXITER=10000;
* G = A*(1-EXP(-K*T))**R;
```

```
PARAMETERS A=0.03 TO 0.13 BY 0.02 K=0.01 TO 0.07 BY 0.01
R=0.005 TO 0.115 BY 0.01;
BOUNDS 0.01 < A < 3, 0.001 < K < 0.1, 0.004 < R < 5;
EKT = EXP(-K*T);
EKT1 = 1 - EKT;
EKTR = (EKT1)**R;
 G = A*EKTR;
 FIT G;
DER.A = EKTR;
DER.K = A*T*R*EKT*EKT1**(R-1);
DER.R = A*EKTR*LOG(EKT1);
RUN;
PROC MODEL DATA=DATA MAXITER=10000;
* V = A*(1-EXP(-K*T))**R;
PARAMETERS A=0.7 TO 2 BY 0.1 K=0.05 R=0.1 TO 5 BY 0.1;
BOUNDS A < 18, 0 < K < 0.2, 0 < R < 20;
EKT = EXP(-K*T);
EKT1 = 1 - EKT;
EKTR = (EKT1)**R;
 V = A*EKTR;
 FIT V;
DER.A = EKTR;
DER.K = A*T*R*EKT*EKT1**(R-1);
DER.R = A*EKTR*LOG(EKT1);
RUN;
PROC GPLOT DATA=DATA;
   PLOT D*T H*T G*T V*T;
RUN;
```

The results of processing the SAS program to fit the
equations for d, h, g, v, with the 5th degree polynomial are shown
in QUADRO 1.

**QUADRO 1.   Results of fitting equations for d, h, g, v,
with the 5th degree polynomial**

The REG Procedure	
Model: MODEL1	
Dependent Variable: d	
Number of Observations Read	150

| Number of Observations Used | 150 | |

Analysis of Variance					
Source	DF	Sum of Squares	Mean Square	F Value	Pr > F
Model	5	10247	2049.44461	279.43	<.0001
Error	144	1056.16737	7.33450		
Corrected Total	149	11303			

Root MSE	2.70823	R-Square	0.9066
Dependent Mean	16.49200	Adj R-Sq	0.9033
Coeff Var	16.42146		

Parameter Estimates					
Variable	DF	Parameter Estimate	Standard Error	t Value	Pr > |t|
The Intercept	1	-1.32542	1.96461	-0.67	0.5010
t	1	0.46030	1.41919	0.32	0.7462
t2	1	0.41648	0.32215	1.29	0.1981
t3	1	-0.04474	0.03072	-1.46	0.1474
t4	1	0.00181	0.00129	1.40	0.1626
t5	1	-0.00002592	0.00001979	-1.31	0.1923

The REG Procedure

Model: MODEL1

Dependent Variable: h

Number of Observations Read	150	
Number of Observations Used	150	

Analysis of Variance					
Source	DF	Sum of Squares	Mean Square	F Value	Pr > F
Model	5	10338	2067.69833	1099.45	<.0001
Error	144	270.81528	1.88066		
Corrected Total	149	10609			

Root MSE	1.37137	R-Square	0.9745
Dependent Mean	14.15733	Adj R-Sq	0.9736
Coeff Var	9.68666		

Parameter Estimates					
Variable	DF	Parameter Estimate	Standard Error	t Value	Pr > |t|
The Intercept	1	0.37657	0.99482	0.38	0.7056
t	1	-0.30491	0.71864	-0.42	0.6720
t2	1	0.32554	0.16313	2.00	0.0479
t3	1	-0.02667	0.01555	-1.71	0.0885
t4	1	0.00094153	0.00065425	1.44	0.1523
t5	1	-0.00001236	0.00001002	-1.23	0.2195

The REG Procedure					
Model: MODEL1					
Dependent Variable: g					

Number of Observations Read	150
Number of Observations Used	150

Analysis of Variance					
Source	DF	Sum of Squares	Mean Square	F Value	Pr > F
Model	5	0.05294	0.01059	131.21	<.0001
Error	144	0.01162	0.00008069		
Corrected Total	149	0.06456			

Root MSE	0.00898	R-Square	0.8200
Dependent Mean	0.02728	Adj R-Sq	0.8138
Coeff Var	32.92769		

Parameter Estimates					
Variable	DF	Parameter Estimate	Standard Error	t Value	Pr > |t|
The Intercept	1	0.00270	0.00652	0.41	0.6789
t	1	-0.00334	0.00471	-0.71	0.4793
t2	1	0.00107	0.00107	1.00	0.3197
t3	1	-0.00007980	0.00010188	-0.78	0.4347
t4	1	0.00000268	0.00000429	0.63	0.5328
t5	1	-3.36043E-8	6.563674E-8	-0.51	0.6095

The REG Procedure					
Model: MODEL1					
Dependent Variable: v					

Number of Observations Read	150
Number of Observations Used	150

Analysis of Variance					
Source	DF	Sum of Squares	Mean Square	F Value	Pr > F
Model	5	11.08361	2.21672	118.90	<.0001
Error	144	2.68461	0.01864		
Corrected Total	149	13.76822			

Root MSE	0.13654	R-Square	0.8050
Dependent Mean	0.29343	Adj R-Sq	0.7982
Coeff Var	46.53290		

Parameter Estimates					
Variable	DF	Parameter Estimate	Standard Error	t Value	Pr > |t|
The Intercept	1	0.01497	0.09905	0.15	0.8801
t	1	-0.01413	0.07155	-0.20	0.8437
t2	1	0.00304	0.01624	0.19	0.8517
t3	1	-0.00008933	0.00155	-0.06	0.9541
t4	1	0.00000321	0.00006514	0.05	0.9608
t5	1	-5.96603E-8	9.976959E-7	-0.06	0.9524

The results of processing the SAS program to fit equations for d, h, g, v, with the Chapman-Richards model are shown in QUADRO 2.

QUADRO 2.   Results of fitting equations for d, h, g, v, with the Chapman-Richards model

The MODEL Procedure							
Nonlinear OLS Summary of Residual Errors							
Equation	DF Model	DF Error	SSE	MSE	Root MSE	R-Square	Adj R-Sq
d	3	147	1094.1	7.4430	2.7282	0.9032	0.9019

Nonlinear OLS Parameter Estimates				
Parameter	Estimate	Approx Std Err	t Value	Approx Pr > |t|
A	28.15501	1.2961	21.72	<.0001
k	0.127712	0.0198	6.44	<.0001

| r | 1.878104 | 0.2857 | 6.57 | <.0001 | | |

The MODEL Procedure						
Nonlinear OLS Summary of Residual Errors						
Equation	DF Model	DF Error	SSE	MSE	Root MSE	R-Square	Adj R-Sq
h	3	147	275.9	1.8767	1.3699	0.9740	0.9736

Nonlinear OLS Parameter Estimates				
Parameter	Estimate	Approx Std Err	t Value	Approx Pr > |t|
A	36.18866	2.5804	14.02	<.0001
k	0.071639	0.00998	7.18	<.0001
r	1.757809	0.1472	11.94	<.0001

The MODEL Procedure							
Nonlinear OLS Summary of Residual Errors							
Equation	DF Model	DF Error	SSE	MSE	Root MSE	R-Square	Adj R-Sq
g	3	147	0.0117	0.000080	0.00893	0.8185	0.8160

Nonlinear OLS Parameter Estimates				
Parameter	Estimate	Approx Std Err	t Value	Approx Pr > |t|
A	0.080285	0.0183	4.39	<.0001
k	0.080389	0.0313	2.57	0.0112
r	2.45536	0.7097	3.46	0.0007

The MODEL Procedure							
Nonlinear OLS Summary of Residual Errors							
Equation	DF Model	DF Error	SSE	MSE	Root MSE	R-Square	Adj R-Sq
v	3	147	2.6854	0.0183	0.1352	0.8050	0.8023

Nonlinear OLS Parameter Estimates				
Parameter	Estimate	Approx Std Err	t Value	Approx Pr > |t|
A	2.567437	2.3777	1.08	0.2820
k	0.050858	0.0412	1.23	0.2189
r	3.40132	1.4572	2.33	0.0209

12 INTRODUCTION TO SAMPLING

When populations are very large, such as forests, it becomes very difficult and costly, if not impossible, to measure all the individuals that make up the population. So a small part of the population is measured and the values for the whole population are estimated.

A sample is a small representative part of a population chosen according to statistical criteria.

The values of the population are called population parameters, while the values of a sample are called sample statistics. Sample statistics are estimates of population parameters, obtained by sampling.

In sampling, continuous and discrete variables collected from sample units of the population are measured or evaluated, and after summarizing and analyzing them, they are transformed into statistics.

12.1 Variables

Variables are the characteristics of the individuals in the population that can be measured, or enumerated, or that can be parameterized.

Examples:

> Measured variable: tree diameter at 1.3 m height;

> Parameterized variable: average spacing between trees (MS) calculated by:

$$EM = \sqrt{(10000m^2 / N)}$$

Where: MS = average spacing between trees in meters; N = number of trees per hectare (N is an enumerated variable).

The main dendrometric variables and tree attributes obtained from forest inventories :

- ➢ Tree species(s), diameter (d) , bark thickness (e), height (h), basal area (g), form factor (f), volume (v), volume without bark (vs);
- ➢ Crown diameter (CD), height of crown base (hbc), crown projection surface (SPC);
- ➢ Stem height (hf), commercial volume (vc),
- ➢ Stem quality (QF);
- ➢ Wooden assortments.

Some forest stand variables obtained from timber forest inventories are:

- ➢ Location (coordinates) and area of the stand;
- ➢ age (t);
- ➢ average diameter ($\bar{d}$), average basal area ($\bar{g}$), average height ($\bar{h}$);
- ➢ number of trees per hectare (N), basal area per hectare (G), volume per hectare (V), volume without bark per hectare (Vs).

12.2 Precision and Accuracy

Precision refers to sampling error, which is easier to obtain than accuracy. Precision represents the lack of accuracy due to the deviation of the sample from the estimated mean, or standard error of the mean.

Accuracy refers to the deviation of the sample from the population parameter including measurement errors and other non-sampling errors, which is difficult to obtain.

12.2.1 Sampling intensity (SI)

The sampling intensity is the percentage of the population that has been sampled, calculated by:

$$AI = 100 . a / A$$

Where: IA = sampling intensity; a = sum of the areas of all the sampling units; A = area of the forest stand.

The sampling intensity must be sufficient to represent the population.

12.2.2 Sample sufficiency

The number of sampling units to represent a forest population can be determined by the characteristics of the population or by the cost and resources available for sampling.

12.2.2.1 Sample sufficiency due to population characteristics

Sampling sufficiency based on the characteristics of the population takes into account the results of a preliminary sample with a few sampling units, known as a pilot inventory, using the mean, variance and maximum sampling error corresponding to a desired confidence probability for the mean.

The sampling intensity as a function of the mean, variance and confidence probability for the mean in a simple random sampling of an infinite population is calculated by the equation:

$$n = t^2 S^2 / E^2$$

Where: n = sufficient number of sample units; t = Student's t-value for a p probability of confidence for the mean and n-1 degrees of freedom; S^2 = sample variance; E = sampling error calculated by multiplying the error limit in percentage by the sample mean.

Sampling sufficiency as a function of inventory costs is calculated by:

$$n = (Ct - C0) / C1$$

Where: n = sufficient number of sampling units; Ct = total cost of the inventory; C0 = fixed costs of planning, equipment, analysis and drawing up the inventory report; C1 = average cost per sampling unit (displacement + demarcation + measurement).

12.3 Sampling methods

Sampling methods are classified according to the approach to a single sampling unit, referring to the type of sampling unit, which can be:

> Sample unit of fixed area - square, rectangular, circular.

> Variable area sampling unit - Bitterlich sampling unit, nearest tree, transects, Prodan's six trees, etc.

12.4 Sampling systems

Sampling systems refer to the way in which the sampling units are distributed in the forest stand:

> Unrestricted random sampling system;

> Random sampling system with restrictions;

> Systematic sampling system without restrictions;

> Systematic sampling system with restrictions;

> Mixed sampling system.

The sampling system determines the statistical approach to the data, defining how the sample variance will be analyzed.

12.5 Sampling procedures

The sampling procedure refers to the structure of the sampling as a whole, it is the combination of the sampling method and the sampling system. Some common sampling procedures are listed below:

Simple random sampling (SRS) - is the simplest type of sampling with unrestricted random distribution of sampling units in the forest stand area;

Stratified random sampling (SRS) - is random sampling with restrictions, represented by the spatial division of the population into strata according to age, site, species, or other characteristic that identifies one part of the population as different from the others.

Systematic sampling in lines (ASF) - is a type of sampling in which the sampling units are located equidistant in lines, also equidistant from each other that cross the population; it is a type of systematic sampling with restrictions if the lines are considered as independent first-stage sampling units.

Systematic network sampling (ASR) - is sampling in which the sampling units are located at the intersection of horizontal and vertical lines, both equidistant, forming a network over the population; it is a type of systematic sampling that can be considered unrestricted if the lines are not considered as sampling units.

13 SAMPLING FOR CUBING

13.1 Nearest Tree Method

13.1.1 Characteristics of the method

In this method, sampling points are distributed systematically or randomly over the vegetation under study and the tree closest to the sampling point is found.

The sampling unit in this method consists exclusively of the tree closest to the sampling point.

The area of the sampling unit (a) is calculated by squaring the corrected distance (D) from the tree to the sampling point, as follows:

$$a = (\,2 \,.\, D\,)^2$$

Where: a = area of the sampling unit; D = distance from the nearest tree to the sampling point.

The area proportionality factor (F) used to convert the statistics per plot into estimates per hectare is calculated by:

$$F = A\,/\,a$$

Where: F = proportionality factor; A = 10000 m² (1 ha); a = area of the sampling unit in m².

The other calculations are carried out as in the fixed area sampling method.

Figura 54 - Installation of the sampling unit using the nearest tree method.

13.1.2 Advantages, disadvantages and precautions

The advantages of the method include:

- ➢ Easy to apply;
- ➢ No special tools are needed to demarcate the units;
- ➢ It can be applied to any type of vegetation;
- ➢ Allows the choice of a single individual per unit when selecting trees for cubing.

The main disadvantage is the need for a large number of sample units to represent the population, due to the excessive variation between them.

The distance from the sampling point to the center of the trunk at the base of the tree must be correctly measured in order to get the correct area of the sampling unit.

13.2 Sampling for cubing

When sampling for accurate tree cubing, the nearest tree method can be used.

To do this, it is necessary to know the diameter data of the trees measured in the plots. With the tree diameter data from the plots, the smallest (LI_{DAP}) and largest diameters (LS_{DAP}) are determined. The number of diameter classes for the population is then calculated using Sturges' rule for samples containing up to 2000 individuals, or Floriano's rule for large samples, or any other rule the researcher prefers. The number of trees to be sampled must be the same in each tree diameter class.

13.2.1 Calculating the number of diameter classes

There are different methods for calculating the number of diameter classes, including the Sturges method, which is useful for small samples of less than 2000 individuals; for samples with a larger number of individuals, the Floriano method can be used. The number of classes calculated by these methods is a simple indication, and the researcher can decide which number of classes to use depending on their experience. The equations used in the two methods to determine the number of diameter classes (k) to be used are described below:

- ➢ Sturges:

$$k = 1 + 3.3 \,.\, \log N$$

Where: k = number of diameter classes; N = total number of trees listed in the sample; log = base 10 logarithm; Example: consider a sample containing 1500 trees, k = 1 + 3.3 * Log 1500 = ~12 DBH classes;

- ➢ Floriano:

$$k = N^{0,175} \,.\, \Delta d^{0,3}$$

Where: k = number of diameter classes; N = total number of trees listed in the sample; k = number of diameter classes; N = total number of trees listed in the sample; Δd = range of diameters in the sample (=LS -LI$_{DAPDAP}$); example: sample containing 8000 trees, with the largest diameter of 45 cm and the smallest of 10 cm; Δd = 45cm-10cm = 35cm; k = (8000)^{0,175} . (35)^{0,3} = 4.82 . 2.91 = ~14 DBH classes.

Assuming the second case in which the stand is sampled has 8000 trees with the smallest diameter of 10 cm and the largest of 45 cm and that, by calculating the number of classes, 14 classes were obtained, and it was planned to cover 70 trees, then 70 trees / 14 classes = 5 trees per DBH class must be sampled. Consequently, 70 sampling points should be distributed over the forest area and at each sampling point a DBH class is drawn, with 5 trees being randomly drawn for each DBH class, i.e. each sampling point will receive a value from 1 to 14, representing the DBH class of the tree to be sampled at the point. The tree to be sampled at each sampling point will be the one closest to the point with the diameter belonging to the class drawn.

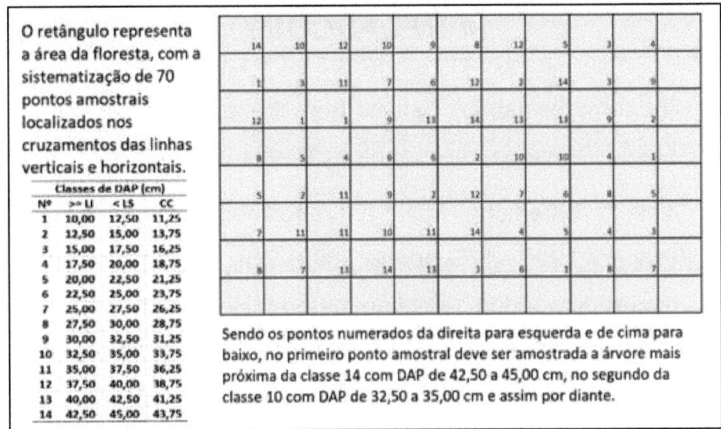

Figura 55 - Systematic distribution of 70 sampling points over the forest area to select trees for rigorous cubing.

14 RELASCOPY

The term Relascopy derives from the equipment known as the Spiegel Relaskop, developed by Walter Bitterlich.

Walter Bitterlich was a renowned forest scientist (19/February 1908 - 9/February 2008) who, in 1947, proposed the angular numbering sampling method. He is one of the best-known personalities in the forestry field and published several scientific papers, including the "sampling by angle counting" method. This revolutionary idea led him to develop the relascope in 1955.

The Mirror Relascope has multiple uses such as:
> Determination of basal area per hectare;
> Measuring tree heights;
> Measuring distances with rangefinders and sights;
> Measurement of trunk diameters at different heights;
> Measuring vertical angles.

The equipment automatically corrects any inclination in the line of sight and is available in four different versions of measurement units.

A manual for using the equipment in Portuguese can be downloaded from the address: <https://www.inventarioflorestal.eu/wp-content/uploads/2012/10/Manual_Relascopio_Telerelascopio.pdf> (BARREIRO et al., 2005).

14.1 Bitterlich sampling method

The statement of the Bitterlich angular sampling method is as follows: "The number of trees (n) in a stand whose diameters at a height of 1.3 m (d) seen from a fixed point appear larger at a given angle (α) is proportional to their basal area per hectare (G)".

14.1.1.1 Method demonstration

Bitterlich developed the method from a bar with an eyepiece 1 meter away to an objective lens 2 centimeters wide (Figura 56). The trees are therefore selected with a probability proportional to the diameter of the trees.

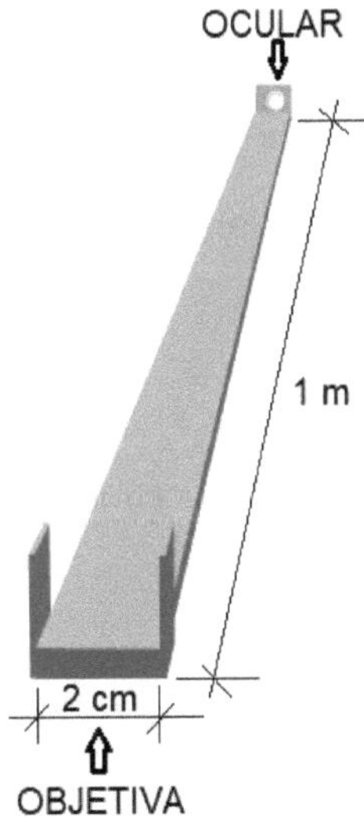

Figura 56 - Bitterlich bar for angle count sampling.

With the Bitterlich bar, you count the trees with a diameter wider than the lens aperture from a point fixed in a 360° rotation. It is suggested that you always start in the north and count clockwise. The number of trees counted (n) is equal to the basal area per hectare (G):

$$G = n$$

Where: G = Basal area per hectare (m²/ha); n = number of trees counted in the sampling unit.

Considering the Figura 57 using the Bitterlich bar, the basal area per hectare is calculated for one tree using the equation:

$$G = K$$

Where: G = Basal area per hectare (m²/ha); K = Basal area factor, also called angular numbering factor.

Since G = n and G = K, G = n . K, then the K value of the Bitterlich bar with an objective aperture of 2% is equal to 1.

The basal area for more than one counted tree is determined by:

$$G = K . n$$

Where: G = Basal area/ha; K = Basal area factor; n = Number of individuals counted at the sampling point with diameters greater than the opening angle (α).

In Figura 57consider the relationship between the measurements:

$$a / L = d / R$$

Where: L = length of the bar; a = angular opening; d = diameter at 1.3 m above the ground; R = radius of the plot.

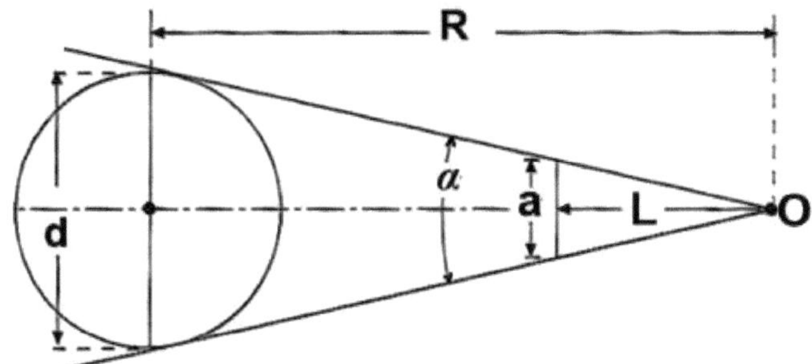

Figura 57 - Variables used to develop the Bitterlich method: α = angle for angle counting; a = width of the Bitterlich bar lens corresponding to the angle α; L = distance from the eyepiece to the bar lens; O = eyepiece; d = tree diameter at 1.3 m height; R = distance from the bar eyepiece to the center of the tree trunk.

If the relationship is true:

$$a/L = d/R_{ii}$$

E, being the surface area of the plot (S):

$$S = \pi \cdot R^2$$

The basal area of a tree (g_i) is calculated as:

$$g_i = \pi \cdot d_i^{\,2} / 4$$

Where: g_i = basal area of tree i; a = lens width (m); L = bar length (m); R_i = distance from the sampling point to the center of the trunk of tree i (m); S = plot area (m²); d_i = diameter of tree i (m).

Therefore, the basal area (g_{UA}) of the sampling unit corresponding to tree "i" is given by:

$$g_{UAi} = g_i / S = (\pi \cdot d_i^{\,2}/4) / \pi \cdot R^2$$

$$g_{UAi} = (d_i^{\,2}/4) / R^2 = (1/4) \cdot (d_i^{\,2}/R^2)$$

$$g_{UAi} = (1/4) \cdot (d_i / R)^2$$

For 1 hectare (10000m²), multiply g_{UA} by 10000m²:

$$G_{UAi} = 10000 \cdot g_{UAi}, \text{ or}$$

$$G_{UAi} = 10^4 \cdot (1/4) \cdot (d_i /R)^2$$

The slope of the terrain means that the horizontal distance to the diameter of the trees at a height of 1.3 m has to be calculated, as the distance is the side of a right triangle and the distance to the tree is the hypotenuse of this triangle.

Based on this principle, in 1955, Bitterlich developed the apparatus known as the Relascope (Figura 58).

Relascope automatically corrects the slope in the line of sight, making it possible to find the correct basal area of the stand in square meters per hectare regardless of the slope of the terrain.

Figura 58 -Relascope with Bitterlich mirror.

Currently, there is a choice of electronic devices with the same functionality as the Relascope, such as the Criterion dendrometer (Figura 59).

Figura 59 - Criterion dendrometer.

14.1.1.2 Angular Numbering Factor, or instrumental constant (K)

Considering the occurrence of a single tree larger than the angle (α) in the sampling unit (N=1), and making:

$$K = GUA_i$$

We obtain the value of the angle numbering factor (K), which is given by:

$$K = 10000. (1/4).(d /R_{ii})^2$$

The basal area per hectare (G) is calculated as the number of trees per hectare (N) multiplied by the average basal area of the trees ($\bar{g}$) of the sampling unit, as:

$$G = N . \bar{g}$$

Continuing analogously:

$$G = N . GUA_i$$

By substituting GUAi, the Bitterlich equation is obtained:

$$G = N . K$$

It is understood that there is a critical radius (R_i) for each tree at the sampling point to be counted, as it has a diameter (d_i) greater than the opening of the angle corresponding to a certain numbering factor (K), calculated by:

$$K = 10000. (1/4).(d /R_{ii})^2$$

Or:

$$K = 2500 (d_i /R_i)^2, R_i = d_i \sqrt{(2500/K)}, d_i = R_i / \sqrt{(2500/K)}$$

14.1.1.3 Doubtful trees

Doubtful trees have been treated in three different ways in angle count sampling:

> By measuring the horizontal distance from the center of the AU to the center of the tree trunk at 1.3 m height, the limit diameter is calculated by d_{lim} =R/$\sqrt{(2500/K)}$; if d_{lim} is smaller than the tree's DBH it is included in the AU; this is the most correct way (Tabela 14);

> Sorting the doubtful trees and excluding the 1st, including the 2nd, and so on;

➢ Counting doubtful trees as 0.5 instead of 1 to calculate G=N.K; this is the worst method and complicates the calculation of averages, as the average has to be calculated using half the diameter or height of the trees.

TABELA 14 - Examples of trees in a sampling unit that should or should not be counted depending on the band (numbering factor), the distance to the center of the sampling unit and the center of the tree trunk at a height of 1.3m .

Tree	K	d (cm)	Distance to the tree (m)	Critical radius (m)	Count? (yes/no)
1	1	28	7,0	14,0	Yes
2	2	28	7,0	7,0	No
3	3	28	7,0	4,7	No
4	4	28	7,0	3,5	No
5	1	35	8,0	17,5	Yes
6	2	35	8,0	8,8	Yes
7	3	35	8,0	5,8	No
8	4	35	8,0	4,4	No

Where: K=instrumental constant; d = tree diameter at 1.3m height; Critical radius = the tree should be excluded from the sampling unit if the distance from the center of its trunk at 1.3m height to the center of the sampling unit is equal to or greater than the critical radius.

14.1.1.4 Sampling by angular numbering test

First, the center of the AU is marked. Starting in a clockwise direction from the north, count the trees with a DBH wider than the chosen angle in a 360° turn. Doubtful trees such as No. 4 should have their doubts resolved by the equation:

$$d_{LIM} = R / [\, 50 \sqrt{(1/K)} \,]$$

Where: d_{LIM} =limit diameter in m; R=distance from the center of the AU to the tree in m; K=angle numbering factor.

If the calculated d_{Lim} is equal to or greater than the tree's measured d, the tree is excluded.

In Figura 60 trees 1 and 8 are outside the AU; tree 4 is doubtful, with d=35cm at a distance of 10 m it is inside because the diameter of the tree is greater than d_{LIM} [d_{LIM} = R / [50 √(1/K)] = 10 / [50 √ (1/3)] = 0.346 m]; 7 trees were counted in the AU with a numbering factor K=3, resulting in 21 m²/ha of basal area.

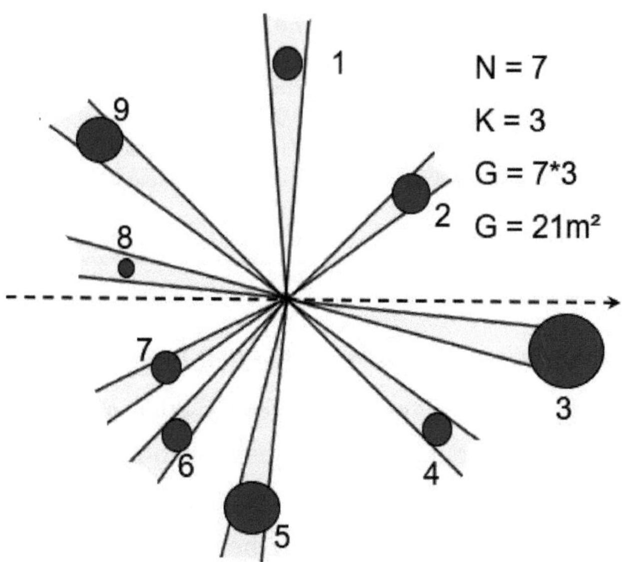

Figura 60 - Bitterlich sampling.

14.2 Statistics in Bitterlich sampling

14.2.1 Average basal area per hectare

The basal area per hectare (G) in m²/ha is obtained directly by multiplying the numbering factor (K) by the number of trees (n) included in the sampling unit:

$$G = K \cdot n$$

14.2.2 Arithmetic mean diameter ($\bar{d}$)

It is calculated using the diameters weighted by the basal area of the trees, using the equation:

$$\bar{d} = \frac{\sum_{i=1}^{n} \dfrac{d_i}{g_i}}{\sum_{i=1}^{n} \dfrac{1}{g_i}}$$

Where: $\bar{d}$ = average diameter of the sampling unit; n = number of trees selected at the sampling point; di = individual diameter of tree "i"; g_i = basal area of tree "i" (m²).

14.2.3 Arithmetic mean height ($\bar{h}$)

It is calculated using the heights weighted by the basal area of the trees, using the equation:

$$\bar{h} = \frac{\sum_{i=1}^{n} \dfrac{h_i}{g_i}}{\sum_{i=1}^{n} \dfrac{1}{g_i}}$$

Where: $\bar{h}$ = average height of the sampling unit; n = number of trees selected at the sampling point; hi = individual height of tree "i"; g_i = basal area of tree "i" (m²).

14.2.4 Frequency per hectare (N)

The Frequency (N), or number of trees per hectare, is obtained from the equation:

$$N = K. \sum_{i=1}^{n} \frac{1}{g_i}$$

Where: N = frequency of trees per hectare; K = basal area factor; gi = basal area of tree "i" (m²).

14.2.5 Volume per hectare (V)

The volume per hectare is determined using the equation:

$$V = K. \sum_{i=1}^{n} \frac{v_i}{g_i}$$

Where: V = Volume per hectare (m³/ha); K = basal area factor; n = number of trees selected at the sampling point; v_i = volume of tree "i" selected (m³); g_i = basal area of tree "i" selected (m²).

14.3 Advantages of angle number sampling

- ➤ Obtain the basal area per hectare immediately;
- ➤ Quick volume calculation when the average form factor is known;
- ➤ Ease and speed of locating the sampling unit (point).

14.4 Disadvantages of sampling by angle numbering

> Usually requires a greater number of units than fixed-area units;

> Higher cost than the fixed area method, in general, for the same accuracy;

> Special equipment required;

> Less accuracy in estimating volume;

> The trees measured may not be the same on two occasions in a continuous inventory, making them unsuitable for growth studies.

14.5 Pay attention to sampling by angle numbering

For greater precision, the distances from the center to the middle of the trunk at the base of the doubtful trees should be measured to determine whether they are inside or outside the sampling unit.

15 REFERENCES

ALVES, A. A. M. **Técnicas de produção florestal**. Lisbon: National Institute for Scientific Research, 1982. 334p.

ARX, G. **Tree-ring research**: much more than just counting rings. Zurich: Swiss Federal Institute for Forest, Snow and Landscape Research WSL. Available at: <https://www.wsl.ch/en/forest/tree-ring-research/tree-ring-research-much-more-than-just-counting-rings.html>. Accessed in: 2021.

BARREIRO, S.; TOMÉ, M.; LUIS, M. **User manuals:** Bitterlich mirror relascope and Telerelascope. Lisbon: UTL/ISA-DEF, GIMREF Technical Report, PT 2/2004.

BATISTA, J. L. F.; COUTO, H. T. Z. The Stereo. **METRVM**, ESALQ, Piracicaba, n.2, October 2002. Available at: <http://cmq.esalq.usp.br/Philodendros/lib/exe/fetch.php?media=lcf0410:biblio:batista-couto-2002.pdf>. Accessed in: 2021.

BIOLOGY ONLINE. **Growth and plant hormones**. Available at: <https://www.biologyonline.com/tutorials/growth-and-plant-hormones>. Accessed in: 2021.

BOURDO Jr., E. A. **The Illustrated book of trees**. London Salamander Books, 2001.

BRACK, C. L.; WOOD, G. B. **Forest Mensuration**: measuring trees, stands and forests for effective forest management. Canberra: Australian National University, 1997. Available at: <https://fennerschool-associated.anu.edu.au/mensuration/BrackandWood1998/MENSHOME.HTM>.

BURKHART, H. E.; TOMÉ, M. **Modeling Forest trees and stands**. Spring, 2012.

COCHRAN, W. G. **Sampling Techniques**. Rio de Janeiro: USAID / Fundo de Cultura, 1965. 555p. Translation by Fernando A. Moreira Barbosa: "COCHRAN, W. G. Sampling techniques, 2ed. John Wiley & Sons, 1953."

DRUSZCZ, J. P. et al. Efficiency of forest inventory with Bitterlich point sampling and line conglomerate in a *Pinus taeda* plantation. **FLORESTA**, Curitiba, PR, v. 42, n. 3, p. 527 - 538, jul./set. 2012.

MICHAELIS ONLINE DICTIONARY. Available at: <https://michaelis.uol.com.br>. Accessed in: 2021.

ESALQ - Luiz de Queiroz College of Agriculture. ESALQ trails: Useful trees. [Piracicaba]: USP/CIAGRI/ESALQ, [1997]. Available at: <http://www.esalq.usp.br/ trilhas/uteis/>. Accessed on: 21/06/2008.

ETH. **Crossdating**. Zurich: ETH Zurich, Tree-ring |Lab. https://fe.ethz.ch/en/research/ dendrochronology-and-phenology/tree-ring-lab/crossdating.html>. Accessed in: 2021.

FAO. **Forest resources assessment working paper 180**: terms and definitions. Rome, 2015.

FLORIANO, E. P. **Forest Phytosociology**. São Gabriel: ed. by the author, 2014.

FLORIANO, E. P. **Forest Management**: towards sustainability and excellence. Rio Largo: ed. by the author, 2018.

GARCIA, P. O.; LOBO-FARIA, P. C. **Metodologias para Levantamentos da Biodiversidade Brasileira**. Juiz de Fora: UFJF/PEGCOL, [2007]. 22p.

GATTO, A. et al. Carbon stocks in soil and biomass in eucalyptus plantations. **Revista Brasileira de Ciência do Solo**, n.34, p.1069-1079, 2010.

GONÇALVES, G. V. **Dendrochronology**: theoretical principles, practical problems and applicability. Évora: CIDEHUS, University of Évora, 2007. Available at: <http://www.alt-shn.org/uploads/3/1/9/2/3192138/dendrocronologia_-_principios_teoricos.pdf>. Accessed in: 2021.

HIGA, R. C. V. et al. **Protocol for measuring and estimating forest biomass and carbon**. Colombo: Embrapa Florestas, 2014. Available at: < https://ainfo.cnptia.embrapa.br/digital/ bitstream/item/121558/1/Doc.-266.pdf>. Accessed in: 2021.

IMAÑA, J. E.; SILVA, G. F.; TICCHETTI, I. Dendrometric variables. **CTF**, Brasília, UNB, v. 4, n. 1, October-2002.

IMAÑA ENCINAS, J; SILVA, G.F.; PINTO, J.R.R. **Age and growth of trees**. CTF, UNB, Brasília, v.7, n.1, 2005.

IUFRO. **The standardization of symbols in forest mensuration**. Maine Agricultural Experiment Station, Technical Bulletin, n.15, 1965. Reprint of the 1959 original.

KRAMER, P. J.; KOZLOWSKI, T. T. **Fisiologia das árvores**. Lisbon: Calouste Gulbenkian Foundation, 1972. 745p.

LAAR, A.; AKÇA, A. **Forest Mensuration**. Springer, 2007.

LTRR. **Andrew E Douglass**: Father of Dendrochronology. Tucson: College of Science, Laboratory of Tree-Ring Research. Available at: <https://www.ltrr.arizona.edu/~cbaisan/Vermont/Erica/AED.pdf>. Accessed in: 2021.

MÜLLER, I. **Trunk shape and wood assortments for *Eucalyptus grandis* Hill ex Maiden, managed at high stem, in the Southeast Region of the State of Rio Grande do Sul**. Thesis (Doctorate in Forest Engineering) - UFSM PPGEF, Santa Maria, 2004.

OLIVEIRA, P. P. A. **Protocol for quantifying soil carbon stocks from the Pecus research network**. São Carlos, SP: Embrapa Pecuária Sudeste, 2014. Available at: <https://www.infoteca.cnptia.embrapa.br/infoteca/bitstream/doc/1006926/1/Documentos116.pdf>. Accessed in: 2021.

ORMOND, J. G. P. **Glossary of Terms Used in Agriculture, Forestry and Environmental Sciences**, 3 ed. Rio de Janeiro: BNDES, 2006.

PÉLLICO NETTO, S.; BRENA, D. A. **Inventário Florestal**. Curitiba, 1997. 316 p.

PRODAN, M. **Mensura forestal**. San José: Inter-American Institute for Cooperation on Agriculture (IICA), 1997. 562 p.

PUNCHES, J. **Tree Growth, Forest Management, and Their Implications for Wood Quality**. Douglas County: Oregon State University, PNW576, September 2004.

RETSLAFF, F. A. S. et al. Conglomerate sampling by the Bitterlich method in Mixed Ombrophilous Forest. **Nativa**, Sinop, v. 02, n. 04, p. 194-198, Oct./Dec. 2014.

SAS. **A simple regression model with correction of heteroscedasticity**. Cary: SAS Institute, 2004.

SAHIN, D. Tracing footprints of environmental events in tree ring chemistry using neutron activation analysis. Doctoral thesis - Ph.D. in Nuclear Engineering, The Pennsylvania State University, 2012. Available at: <https://www.researchgate.net/publication/265260268>.

SANTOS, C. F. et al. Biomass in *Pinus elliottii* Engelm: a drain for carbon. **Revista de Engenharia e Tecnologia**, v.11, n.1, p.50-55, Apr/2019.

SCHNEIDER, P. R. **Regression analysis applied to forestry engineering**. Santa Maria: UFSM, CEPEF, 1998.

SFB. **Forest Stock - Biomass - Tables and Graphs**. Brasília: Brazilian Forest Service, National Forest Information System, 2016. Available at: <https://snif.florestal.gov.br/pt-br/estoques-das-florestas/624-tabelas-e-graficos?tipo=tableau&modal=1>. Accessed on: 2024.

SFB. **Forest Stock - Carbon**. Brasília: Brazilian Forest Service, National Forest Information System, 2020. Available at: <https://snif.florestal.gov.br/pt-br/estoques-das-florestas/623-estoque-das-florestas-carbono>. Accessed in: 2024.

SILVA, C. A. Carbon stock in forest aerial biomass in commercial plantations of *Eucalyptus spp.* **Scientia Forestalis**, Piracicaba, v. 43, n. 105, p. 135-146, mar/2015.

SILVA, J.A.A; PAULA NETO, F. **Princípios básicos de dendrometria**. Recife: UFRPE, 1979.

SILVESTRE, R.; BONAZZA, M.; STANG, M.; LIMA, G.C.P.; KOEPSEL, D.A.; MARCO, F.T.; CIARNOSCHI, L.D.; SCARIOT, R.; MORÊS, D.F. Volumetric equations in Pinus taeda L. stands in the municipality of Lages-SC. **Nativa**, Sinop, v. 02, n. 01, p. 01-05, jan./mar. 2014.

SIT, V. **Catalog of curves for curve fitting**. Biometrics information handbook series, Victoria (Canada), Forest Science Research Branch, n.4, 1994. ISSN 1183-9759.

STORCK, L.; LOPES, S. J. **Experimentation II**. Santa Maria: UFSM, CCR/Dep. Fitotecnia, 1998. 205 p.

BRUCHEZ, T. Like an Onion, Trees Have Layers. Princeton: Tree service experts, 2017. Available at: <http://www.treeservice.expert/like-an-onion-trees-have-layers>.

PEARSON, T.; WALKER, S.; BROWN, S. **Sourcebook for land use, land-use change and forestry projects**. Washington DC: The World Bank, 2005. Available at: <https://winrock.org/wp-content/uploads/2016/03/Winrock-BioCarbon_Fund_Sourcebook-compressed.pdf>. Accessed on: 2024.

USP. **Introduction to plant biology**. São Carlos, 2002. Available at: <biologia.ifsc.usp.br/bio3/outros/02-Morfologia.pdf>. Accessed on: 9/11/2014.

WEST, P.W. **Tree and forest measurement**, 2ed. Berlin: Springer-Verlag, 2009.

TERRAGES. Sales website. Campo Maior, Portugal, 2021. Available at: <https://www.terrages.pt/>. Accessed in 2021.

THEISEN, J. Wooster Tree Ring Lab Ready for Business. Wooster, USA: Wooster Geologists. Department of Geology, College of Wooster, June 10, 2011. Available at: <https://woostergeologists.scotblogs.wooster.edu/2011/06/10/wooster-tree-ring-lab-ready-for-business/>.

TOMÉ, M. **Inventory of forest resources**. Lisbon: Universidade Técnica de Lisboa / ISA / GIMREF, Textos pedagógicos, n. 1, 2002.

VANDERZANDEN, A. M. **How hormones and growth regulators affect your plants**. OSU Extension Service, July 2012. Available at: <https://extension.oregonstate.edu/gardening /techniques/how-hormones-growth-regulators-affect-your-plants>. Accessed in: 2021.

WOODTECH. **Logmeter**. Santiago, Chile: Woodtech Measurement Solutions, 2021. Available at: <https://www.woodtechms.com/logmeter>. Accessed: 2021.

VANTEC. **Laminating lathe 1400**. Available at: <https://www.youtube.com/watch?v=ilu5i Udjyw4>. Accessed: 2021.

16 Appendix A
Trunk analysis - sampling

TABELA 15 - Actual volumes of the trunk sections of 5 14-year-old *Eucalyptus urograndis* trees calculated using the Smalian method

Tree no.	Section no.	d (cm)	h (m)	$n_{j,i}$	L_i (m)	h_i (m)	h/h_i	d_i (cm)	g_i (m²)	v_i (m³)	v (m³)
1	0	32,4	37,80	0,0	0,15	0,15	0,0040	36,30	0,00285	0,00043	0,00043
1	1	32,4	37,80	1.1	2,75	2,90	0,0767	29,50	0,00232	0,00711	
1	2	32,4	37,80	1.2	2,70	5,60	0,1481	27,20	0,00214	0,00601	
1	3	32,4	37,80	1.3	2,70	8,30	0,2196	26,40	0,00207	0,00568	
1	4	32,4	37,80	1.4	2,70	11,00	0,2910	24,00	0,00188	0,00534	
1	5	32,4	37,80	1.5	2,70	13,70	0,3624	22,80	0,00179	0,00496	
1	6	32,4	37,80	1.6	2,70	16,40	0,4339	20,70	0,00163	0,00461	
1	7	32,4	37,80	1.7	2,70	19,10	0,5053	20,00	0,00157	0,00432	
1	8	32,4	37,80	1.8	2,70	21,80	0,5767	17,00	0,00134	0,00392	0,04196
1	9	32,4	37,80	2.1	2,70	24,50	0,6481	15,30	0,00120	0,00342	
1	10	32,4	37,80	2.2	2,00	26,50	0,7011	13,50	0,00106	0,00226	
1	11	32,4	37,80	2.3	2,00	28,50	0,7540	12,40	0,00097	0,00203	
1	12	32,4	37,80	2.4	2,00	30,50	0,8069	10,30	0,00081	0,00178	0,00950
1	13	32,4	37,80	3,0	7,30	32,50	0,8598	0,00	0,00000	0,00197	0,00197
2	0	15,2	27,60	0,0	0,10	0,10	0,0036	18,20	0,00143	0,00014	0,00014
2	1	15,2	27,60	2.1	2,05	2,15	0,0779	14,50	0,00114	0,00263	
2	2	15,2	27,60	2.2	2,00	4,15	0,1504	13,50	0,00106	0,00220	
2	3	15,2	27,60	2.3	2,00	6,15	0,2228	12,60	0,00099	0,00205	

Tree no.	Section no.	d (cm)	h (m)	$n_{j,i}$	L_i (m)	h_i (m)	h/h_i	d_i (cm)	g_i (m²)	v_i (m³)	v (m³)
2	4	15,2	27,60	2.4	2,00	8,15	0,2953	12,00	0,00094	0,00193	
2	5	15,2	27,60	2.5	2,00	10,15	0,3678	11,10	0,00087	0,00181	
2	6	15,2	27,60	2.6	2,00	12,15	0,4402	10,20	0,00080	0,00167	
2	7	15,2	27,60	2.7	2,00	14,15	0,5127	9,30	0,00073	0,00153	
2	8	15,2	27,60	2.8	2,00	16,15	0,5851	8,20	0,00064	0,00137	
2	9	15,2	27,60	2.9	2,00	18,15	0,6576	7,50	0,00059	0,00123	
2	10	15,2	27,60	2.10	2,00	20,15	0,7301	6,00	0,00047	0,00106	
2	11	15,2	27,60	2.11	2,00	22,15	0,8025	4,80	0,00038	0,00085	0,01835
2	12	15,2	27,60	3,0	5,45	27,60	1,0000	0,00	0,00000	0,00068	0,00068
3	0	19,8	32,00	0,0	0,15	0,15	0,0047	26,20	0,00206	0,00031	0,00031
3	1	19,8	32,00	1.1	2,75	2,90	0,0906	18,90	0,00148	0,00487	
3	2	19,8	32,00	2.1	2,70	5,60	0,1750	17,60	0,00138	0,00387	
3	3	19,8	32,00	2.2	2,70	8,30	0,2594	16,30	0,00128	0,00359	
3	4	19,8	32,00	2.3	2,70	11,00	0,3438	15,00	0,00118	0,00332	0,01565
3	5	19,8	32,00	2.4	2,00	13,00	0,4063	13,90	0,00109	0,00227	
3	6	19,8	32,00	2.5	2,00	15,00	0,4688	13,10	0,00103	0,00212	
3	7	19,8	32,00	2.6	2,00	17,00	0,5313	12,40	0,00097	0,00200	
3	8	19,8	32,00	2.7	2,00	19,00	0,5938	11,30	0,00089	0,00186	
3	9	19,8	32,00	2.8	2,00	21,00	0,6563	10,20	0,00080	0,00169	
3	10	19,8	32,00	2.9	2,00	23,00	0,7188	8,70	0,00068	0,00148	
3	11	19,8	32,00	2.10	2,00	25,00	0,7813	7,60	0,00060	0,00128	
3	12	19,8	32,00	2.11	2,00	27,00	0,8438	6,20	0,00049	0,00108	0,01379

Tree no.	Section no.	d (cm)	h (m)	$n_{j,i}$	L_i (m)	h_i (m)	h/h_i	d_i (cm)	g_i (m²)	v_i (m³)	v (m³)
3	13	19,8	32,00	3,0	5,00	32,00	1,0000	0,00	0,00000	0,00081	0,00081
4	0	22,7	33,70	0,0	0,15	0,15	0,0045	27,00	0,00212	0,00032	0,00032
4	1	22,7	33,70	1.1	2,70	2,90	0,0861	22,30	0,00175	0,00523	
4	2	22,7	33,70	1.2	2,70	5,60	0,1662	21,20	0,00167	0,00461	
4	3	22,7	33,70	1.3	2,70	8,30	0,2463	19,70	0,00155	0,00434	
4	4	22,7	33,70	2.1	2,70	11,00	0,3264	18,50	0,00145	0,00405	
4	5	22,7	33,70	2.2	2,70	13,70	0,4065	16,20	0,00127	0,00368	0,02191
4	6	22,7	33,70	2.3	2,00	16,40	0,4866	15,00	0,00118	0,00245	
4	7	22,7	33,70	2.4	2,00	18,40	0,5460	14,20	0,00112	0,00229	
4	8	22,7	33,70	2.5	2,00	20,40	0,6053	13,30	0,00104	0,00216	
4	9	22,7	33,70	2.6	2,00	22,40	0,6647	12,00	0,00094	0,00199	
4	10	22,7	33,70	2.7	2,00	24,40	0,7240	10,40	0,00082	0,00176	
4	11	22,7	33,70	2.8	2,00	26,40	0,7834	9,30	0,00073	0,00155	
4	12	22,7	33,70	2.9	2,00	28,40	0,8427	7,30	0,00057	0,00130	
4	13	22,7	33,70	2.10	2,00	30,40	0,9021	5,30	0,00042	0,00099	0,01449
4	14	22,7	33,70	3,0	3,30	33,70	1,0000	0,00	0,00000	0,00046	0,00046
5	0	27,2	34,30	0,0	0,10	0,10	0,0029	33,80	0,00265	0,00027	0,00027
5	1	27,2	34,30	1.1	2,70	2,85	0,0831	25,60	0,00201	0,00630	
5	2	27,2	34,30	1.2	2,70	5,55	0,1618	23,60	0,00185	0,00522	
5	3	27,2	34,30	1.3	2,70	8,25	0,2405	22,00	0,00173	0,00483	
5	4	27,2	34,30	1.4	2,70	10,95	0,3192	20,20	0,00159	0,00447	
5	5	27,2	34,30	1.5	2,70	13,65	0,3980	18,50	0,00145	0,00410	0,02493

Tree no.	Section no.	d (cm)	h (m)	$n_{j,i}$	L_i (m)	h_i (m)	h/h_i	d_i (cm)	g_i (m²)	v_i (m³)	v (m³)
5	6	27,2	34,30	2.1	2,00	16,35	0,4767	17,00	0,00134	0,00279	
5	7	27,2	34,30	2.2	2,00	19,05	0,5554	15,20	0,00119	0,00253	
5	8	27,2	34,30	2.3	2,00	21,05	0,6137	13,70	0,00108	0,00227	
5	9	27,2	34,30	2.4	2,00	23,05	0,6720	12,60	0,00099	0,00207	
5	10	27,2	34,30	2.5	2,00	25,05	0,7303	11,30	0,00089	0,00188	
5	11	27,2	34,30	2.6	2,00	27,05	0,7886	9,20	0,00072	0,00161	
5	12	27,2	34,30	2.7	2,00	29,05	0,8469	7,00	0,00055	0,00127	0,01441
5	13	27,2	34,30	3,0	5,25	34,30	1,0000	0,00	0,00000	0,00096	0,00096

Where: d = tree diameter; h = tree height; $n_{j,i}$ = order number of section i of assortment j; L_i = length of section i; h_i = height of section i; d_i = diameter of section i; g_i = cross-sectional area of section i; v_i = volume of section i.

17 Appendix B
Stem analysis - estimates

TABELA 16 - Estimated volumes of the trunk sections of 5 14-year-old *Eucalyptus urograndis* trees calculated using the Smalian method

Tree no.	Section no.	d (cm)	h (m)	$n_{j,i}$	Li (m)	hi (m)	hi/h	di (cm)	gi (m²)	vi (m³)	v (m³)
1	0	32,4	37,80	0,0	0,15	0,15	0,0040	39,03	0,00307	0,00046	0,00046
1	1	32,4	37,80	1.1	2,70	2,85	0,0754	32,07	0,00252	0,00754	
1	2	32,4	37,80	1.2	2,70	5,55	0,1468	28,53	0,00224	0,00643	
1	3	32,4	37,80	1.3	2,70	8,25	0,2183	26,67	0,00209	0,00585	
1	4	32,4	37,80	1.4	2,70	10,95	0,2897	25,37	0,00199	0,00552	
1	5	32,4	37,80	1.5	2,70	13,65	0,3611	23,99	0,00188	0,00523	
1	6	32,4	37,80	1.6	2,70	16,35	0,4325	22,30	0,00175	0,00491	
1	7	32,4	37,80	1.7	2,70	19,05	0,5040	20,30	0,00159	0,00452	
1	8	32,4	37,80	1.8	2,70	21,75	0,5754	18,16	0,00143	0,00408	
1	9	32,4	37,80	1.9	2,70	24,45	0,6468	16,01	0,00126	0,00362	0,04769
1	10	32,4	37,80	2.2	2,00	26,45	0,6997	14,47	0,00114	0,00239	
1	11	32,4	37,80	2.3	2,00	28,45	0,7526	12,90	0,00101	0,00215	
1	12	32,4	37,80	2.4	2,00	30,45	0,8056	11,15	0,00088	0,00189	0,00643
1	13	32,4	37,80	3,0	7,35	37,80	1,0000	0,00	0,00000	**0,00215**	0,00215
2	0	15,2	27,60	0,0	0,10	0,10	0,0036	18,33	0,00144	0,00014	0,00014
2	1	15,2	27,60	2.1	2,00	2,10	0,0761	15,02	0,00118	0,00262	
2	2	15,2	27,60	2.2	2,00	4,10	0,1486	13,36	0,00105	0,00223	
2	3	15,2	27,60	2.3	2,00	6,10	0,2210	12,49	0,00098	0,00203	

Tree no.	Section no.	d (cm)	h (m)	$n_{j,i}$	Li (m)	hi (m)	hi/h	di (cm)	gi (m²)	vi (m³)	v (m³)
2	4	15,2	27,60	2.4	2,00	8,10	0,2935	11,87	0,00093	0,00191	
2	5	15,2	27,60	2.5	2,00	10,10	0,3659	11,21	0,00088	0,00181	
2	6	15,2	27,60	2.6	2,00	12,10	0,4384	10,39	0,00082	0,00170	
2	7	15,2	27,60	2.7	2,00	14,10	0,5109	9,43	0,00074	0,00156	
2	8	15,2	27,60	2.8	2,00	16,10	0,5833	8,41	0,00066	0,00140	
2	9	15,2	27,60	2.9	2,00	18,10	0,6558	7,39	0,00058	0,00124	
2	10	15,2	27,60	2.10	2,00	20,10	0,7283	6,40	0,00050	0,00108	
2	11	15,2	27,60	2.11	2,00	22,10	0,8007	5,31	0,00042	0,00092	0,01850
2	12	15,2	27,60	3,0	5,50	27,60	1,0000	0,00	0,00000	0,00077	0,00077
3	0	19,8	32,00	0,0	0,15	0,15	0,0047	23,79	0,00187	0,00028	0,00028
3	1	19,8	32,00	1.1	2,70	2,85	0,0891	19,06	0,00150	0,00454	
3	2	19,8	32,00	2.1	2,70	5,55	0,1734	16,94	0,00133	0,00382	
3	3	19,8	32,00	2.2	2,70	8,25	0,2578	15,85	0,00124	0,00348	0,01184
3	4	19,8	32,00	2.3	2,00	10,25	0,3203	15,16	0,00119	0,00244	
3	5	19,8	32,00	2.4	2,00	12,25	0,3828	14,37	0,00113	0,00232	
3	6	19,8	32,00	2.5	2,00	14,25	0,4453	13,42	0,00105	0,00218	
3	7	19,8	32,00	2.6	2,00	16,25	0,5078	12,34	0,00097	0,00202	
3	8	19,8	32,00	2.7	2,00	18,25	0,5703	11,19	0,00088	0,00185	
3	9	19,8	32,00	2.8	2,00	20,25	0,6328	10,04	0,00079	0,00167	
3	10	19,8	32,00	2.9	2,00	22,25	0,6953	8,92	0,00070	0,00149	
3	11	19,8	32,00	2.10	2,00	24,25	0,7578	7,79	0,00061	0,00131	
3	12	19,8	32,00	2.11	2,00	26,25	0,8203	6,48	0,00051	0,00112	0,01640

Tree no.	Section no.	d (cm)	h (m)	$n_{j,i}$	Li (m)	hi (m)	hi/h	di (cm)	gi (m²)	vi (m³)	v (m³)
3	13	19,8	32,00	3,0	5,75	32,00	1,000	0,00	0,00000	0,00098	0,00098
4	0	22,7	33,70	0,0	0,15	0,15	0,0045	27,30	0,00214	0,00032	0,00032
4	1	22,7	33,70	1.1	2,70	2,85	0,0846	22,05	0,00173	0,00523	
4	2	22,7	33,70	1.2	2,70	5,55	0,1647	19,59	0,00154	0,00441	
4	3	22,7	33,70	1.3	2,70	8,25	0,2448	18,33	0,00144	0,00402	
4	4	22,7	33,70	2.1	2,70	10,95	0,3249	17,32	0,00136	0,00378	
4	5	22,7	33,70	2.2	2,70	13,65	0,4050	16,11	0,00127	0,00354	0,02099
4	6	22,7	33,70	2.3	2,00	15,65	0,4644	15,02	0,00118	0,00244	
4	7	22,7	33,70	2.4	2,00	17,65	0,5237	13,81	0,00108	0,00226	
4	8	22,7	33,70	2.5	2,00	19,65	0,5831	12,56	0,00099	0,00207	
4	9	22,7	33,70	2.6	2,00	21,65	0,6424	11,31	0,00089	0,00187	
4	10	22,7	33,70	2.7	2,00	23,65	0,7018	10,09	0,00079	0,00168	
4	11	22,7	33,70	2.8	2,00	25,65	0,7611	8,85	0,00070	0,00149	
4	12	22,7	33,70	2.9	2,00	27,65	0,8205	7,42	0,00058	0,00128	
4	13	22,7	33,70	2.10	2,00	29,65	0,8798	5,49	0,00043	0,00101	0,01412
4	14	22,7	33,70	3,0	4,05	33,70	1,000	0,00	0,00000	0,00058	0,00058
5	0	27,2	34,30	0,0	0,10	0,10	0,0029	32,88	0,00258	0,00026	0,00026
5	1	27,2	34,30	1.1	2,70	2,80	0,0816	26,57	0,00209	0,00630	
5	2	27,2	34,30	1.2	2,70	5,50	0,1603	23,58	0,00185	0,00532	
5	3	27,2	34,30	1.3	2,70	8,20	0,2391	22,06	0,00173	0,00484	
5	4	27,2	34,30	1.4	2,70	10,90	0,3178	20,86	0,00164	0,00455	
5	5	27,2	34,30	1.5	2,70	13,60	0,3965	19,47	0,00153	0,00428	

Tree no.	Section no.	d (cm)	h (m)	$n_{j,i}$	L_i (m)	h_i (m)	h_i/h	d_i (cm)	g_i (m²)	v_i (m³)	v (m³)
5	6	27,2	34,30	2.1	2,70	16,30	0,4752	17,74	0,00139	0,00395	
5	7	27,2	34,30	2.2	2,70	19,00	0,5539	15,79	0,00124	0,00356	0,03279
5	8	27,2	34,30	2.3	2,00	21,00	0,6122	14,31	0,00112	0,00236	
5	9	27,2	34,30	2.4	2,00	23,00	0,6706	12,86	0,00101	0,00213	
5	10	27,2	34,30	2.5	2,00	25,00	0,7289	11,43	0,00090	0,00191	
5	11	27,2	34,30	2.6	2,00	27,00	0,7872	9,90	0,00078	0,00168	
5	12	27,2	34,30	2.7	2,00	29,00	0,8455	8,02	0,00063	0,00141	0,00949
5	13	27,2	34,30	3,0	5,30	34,30	1,0000	0,00	0,00000	0,00111	0,00111

Where: d = tree diameter; h = tree height; $n_{j,i}$ = order number of section i of assortment j; L_i = length of section i; h_i = height of section i; d_i = diameter of section i; g_i = cross-sectional area of section i; v_i = volume of section i.

18 Appendix C
Enrollment in SAS® OnDemand for Academics

1) To sign up for SAS® OnDemand for Academics, go to:

https://www.sas.com/pt_br/software/on-demand-for-academics.html

2) The screen below opens, where you have to click on **Access now**:

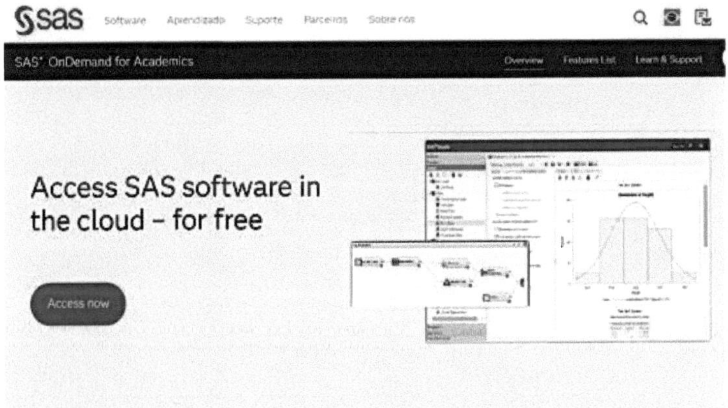

3) The following screen opens, where you have to click on **SAS Profile**:

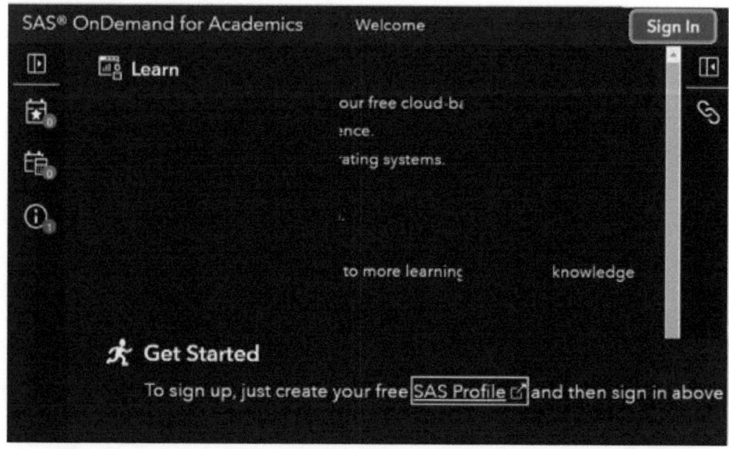

4) Fill in the form on the next page, reproduced below:

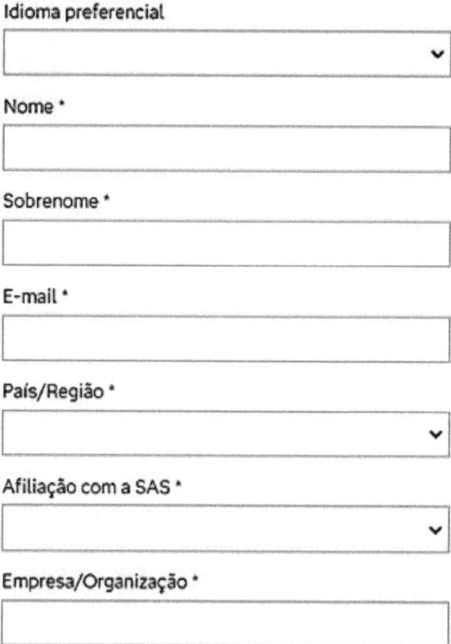

5) On the same page, check **I agree with the terms of use and conditions** and click on **create profile**:

6) And follow the instructions to activate your profile.

19 Appendix D
Diameter and height
data for 60 trees

TABELA 17 - Diameter and height data from 60 trees

Tree	d	h	Tree	d	h	Tree	d	h
1	20.0	25.0	21	20.0	22.0	41	20.0	22.0
2	21.0	27.0	22	21.0	21.0	42	21.0	21.0
3	22.0	28.5	23	22.0	24.2	43	22.0	19.8
4	23.0	30.0	24	23.0	23.0	44	23.0	23.0
5	24.0	31.8	25	24.0	21.6	45	24.0	26.4
6	25.0	33.0	26	25.0	22.5	46	25.0	22.5
7	26.0	34.1	27	26.0	28.6	47	26.0	26.0
8	27.0	35.3	28	27.0	29.7	48	27.0	27.0
9	28.0	36.6	29	28.0	28.0	49	28.0	25.2
10	29.0	37.5	30	29.0	31.9	50	29.0	31.9
11	30.0	38.2	31	30.0	33.0	51	30.0	33.0
12	31.0	39.0	32	31.0	34.1	52	31.0	31.0
13	32.0	39.4	33	32.0	35.2	53	32.0	32.0
14	33.0	40.0	34	33.0	29.7	54	33.0	33.0
15	34.0	40.5	35	34.0	37.4	55	34.0	37.4
16	35.0	41.0	36	35.0	35.0	56	35.0	35.0
17	36.0	41.3	37	36.0	32.4	57	36.0	32.4
18	37.0	41.7	38	37.0	37.0	58	37.0	40.7
19	38.0	42.0	39	38.0	41.8	59	38.0	34.2
20	39.0	42.0	40	39.0	35.1	60	39.0	39.0

yes
I want morebooks!

Buy your books fast and straightforward online - at one of world's fastest growing online book stores! Environmentally sound due to Print-on-Demand technologies.

Buy your books online at
www.morebooks.shop

Kaufen Sie Ihre Bücher schnell und unkompliziert online – auf einer der am schnellsten wachsenden Buchhandelsplattformen weltweit! Dank Print-On-Demand umwelt- und ressourcenschonend produziert.

Bücher schneller online kaufen
www.morebooks.shop

Printed by Books on Demand GmbH, Norderstedt / Germany